Praise for *Wikinomics*

"An intriguing and important book that belongs on your shelf."
—*The Washington Post*

"This dazzlingly timely book is a pretty accurate and comprehensive account of how the [Web 2.0] technological trend is translating into a business trend and offers some ground-floor advice on how to apply 'wiki' thinking to existing businesses." —*Financial Times*

"An interesting and compelling survey of successful collaborative ventures... *Wikinomics* is the right book at the right time and likely portends new openings for the future of businesses big and small."
—*The Miami Herald*

"No company today, no matter how large or how global, can innovate fast enough or big enough by itself.... *Wikinomics* reveals the next historic step—the art and science of mass collaboration where companies open up to the world. It is an important book."
—A. G. Lafley, CEO, Procter & Gamble

"*Wikinomics* heralds the biggest change in collaboration to date. Thanks to the Internet, masses of people outside the boundaries of traditional hierarchies can innovate to produce content, goods, and services. In order to understand the opportunities this presents for companies, read this book." —Eric Schmidt, CEO, Google

"Not only a superb book, but an essential one for anyone who wants to understand the major forces that will revolutionize the way organizations perform and the way they are led." —Warren Bennis

"An insightful and engaging book."
—John Chambers, president and CEO, Cisco Systems

"*Wikinomics* illuminates the truth we are seeing in markets around the globe: the more you share, the more you win. *Wikinomics* sheds light on the many faces of business collaboration and presents a powerful new strategy for business leaders in a world where customers, employees, and low-cost producers are seizing control."

—Brian Fetherstonhaugh, chairman and CEO,
OgilvyOne Worldwide

"A MapQuest–like guide to the emerging business-to-consumer relationship. This book should be invaluable to any manager—helping us chart our way in an increasingly digital world."

—Tony Scott, senior vice president and
chief information officer, The Walt Disney Company

"Knowledge creation happens in social networks where people learn and teach each other. *Wikinomics* shows where this phenomenon is headed when turbocharged to engage the ideas and energy of customers, suppliers, and producers in mass collaboration. It's a must-read for those who want a map of where the world is headed."

—Noel Tichy, professor, University of Michigan and
author of *Cycle of Leadership*

"A deeply profound and hopeful book. *Wikinomics* provides compelling evidence that the emerging 'creative commons' can be a boon, not a threat to business. Every CEO should read this book and heed its wise counsel if they want to succeed in the emerging global economy."

—Klaus Schwab, founder and executive chairman,
World Economic Forum

"Business executives who want to be able to stay competitive in the future should read this compelling and excellently written book."

—Tiffany Olson, president and CEO,
Roche Diagnostics Corporation, North America

"One of the most profound shifts transforming business and society in the early twenty-first century is the rapid emergence of open, collaborative innovation models. *Wikinomics* captures these shifts and explains them in a way that demands the immediate attention of every business, academic and government leader who is interested in driving change."

—Nick Donofrio, EVP of Innovation and Technology,
IBM Corporation

"After the crash of the dot-coms many business leaders breathed a sigh of relief that their traditional business model was safe. *Wikinomics* shows how such complacency is folly. Fueled by the global adoption of the Internet, which has enabled new forms of collaboration and spawned new types of communities, we are entering a second bigger wave of the Internet age. What an insightful analysis of the business upheaval we are facing!"

—Steven L. Sheinheit, EVP and chief information officer, MetLife

"*Wikinomics* highlights a sea change in business today. Customers have more knowledge and power, corporate boundaries are becoming more porous, and people can now collaborate on a scale previously unthinkable. *Wikinomics* will help you understand the changes, why they should be good news for businesses, and how to win in this new world."

—Gordon Nixon, chief executive officer, Royal Bank of Canada

"I love this book. How counterintuitive that openness, peering, sharing, and acting globally would become key to corporate competitiveness, growth and profit. Mass collaboration is the most disruptive development in business in a long time. Consider *Wikinomics* your survival kit."

—Ross Mayfield, CEO, Socialtext

"An extraordinary book. Forward thinking leaders need to understand what this book offers. It was highly meaningful learning for me."

—Michael H. McCain, CEO, Maple Leaf Foods

"Tapscott and Williams have found a way to help companies gain efficiencies by revealing a unique blend of peer interaction, creative modeling and collaboration, or as they call it—wikinomics."

—Ann M. Purr, FLMI, CSP, PCS,
second vice president of information management, LOMA

ABOUT THE AUTHORS

Don Tapscott (left) is an internationally sought consultant on business strategy and organizational transformation. He is the founder of New Paradigm, now owned by nGenera, the leader in enterprise collaboration.

Anthony D. Williams (right) is a senior fellow at the Lisbon Council and a strategic adviser to governments, international institutions, and several Fortune 500 firms.

WIKINOMICS

*How Mass Collaboration
Changes Everything*

EXPANDED EDITION

Don Tapscott
and Anthony D. Williams

PORTFOLIO / PENGUIN

PORTFOLIO / PENGUIN
Published by the Penguin Group
Penguin Group (USA) Inc., 375 Hudson Street, New York, New York 10014, U.S.A.
Penguin Group (Canada), 90 Eglinton Avenue East, Suite 700, Toronto,
Ontario, Canada M4P 2Y3 (a division of Pearson Penguin Canada Inc.)
Penguin Books Ltd, 80 Strand, London WC2R 0RL, England
Penguin Ireland, 25 St Stephen's Green, Dublin 2, Ireland (a division of Penguin Books Ltd)
Penguin Group (Australia), 250 Camberwell Road, Camberwell, Victoria 3124, Australia
(a division of Pearson Australia Group Pty Ltd)
Penguin Books India Pvt Ltd, 11 Community Centre, Panchsheel Park,
New Delhi – 110 017, India
Penguin Group (NZ), 67 Apollo Drive, Rosedale, North Shore 0632, New Zealand
(a division of Pearson New Zealand Ltd)
Penguin Books (South Africa) (Pty) Ltd, 24 Sturdee Avenue, Rosebank, Johannesburg 2196,
South Africa

Penguin Books Ltd, Registered Offices:
80 Strand, London WC2R 0RL, England

Published in the United States of America by Portfolio, a member of Penguin Group (USA)
Inc. 2008
This paperback edition published 2010

While the author has made every effort to provide accurate telephone numbers and Internet
addresses at the time of publication, neither the publisher nor the author assumes any responsibility for errors or for changes that occur after publication. Further, publisher does
not have any control over and does not assume any responsibility for author or third-party
Web sites or their content.

10 9 8 7 6 5 4 3 2 1

ISBN 978-1-59184-193-7 (hc.)
ISBN 978-1-59184-367-2 (pbk.)
CIP data available

Printed in the United States of America
Set in Scala • Designed by Jaime Putorti

We dedicate this book to our children, Alex and Niki Tapscott and Immanuel and Evan Williams. We hope that it helps our generation open up the economy to yours.

PREFACE

At the end of 2006, the same week *Wikinomics* was first published, *Time* magazine chose "YOU," the online collaborator, as its "Person of the Year." *Time* was right to highlight the explosion of social networking. MySpace was growing at two million new registrants per week (and with over two hundred million members was on its way to half a billion). Most college students in the United States were on Facebook. A new blog was created every second of the day, twenty-four hours a day. It seemed that "You" really were changing the world.

At the time we were thrilled to launch our book at such an auspicious moment in history. But looking back on it, it was *so* 2006!

As we explained in *Wikinomics*, the Internet is no longer about hooking up online, creating a gardening community, or putting a video on YouTube. "User-generated media" and "social networking" are really just the tip of an iceberg.

Thanks to an abundance of new tools for collaboration, companies are beginning to conceive, design, develop, and distribute products and services in profoundly new ways. Closed, hierarchical corporations that once innovated in secret can now tap, and contribute to, a much larger global talent pool—one that opens up the world of knowledge workers to every organization seeking a uniquely qualified mind to solve their problem. Smart companies can design and assemble products with their customers, and in some cases customers can do the majority of the value creation.

The evidence continues to mount in support of our assertion that the corporation may be going through the biggest change in its short history.

Indeed, the impact of *Wikinomics* has been quite remarkable. The concept itself has become a word in the vernacular. The word "wikinomics" (small w) is being used in ways and by groups that we never anticipated. Everyone from urban planners to fundamentalist Christians are discussing the concept. There is even a "Wikinomics for Obama" group on my.barackobama.com. In many countries large numbers of people and firms are using wikinomics to form communities and innovate collaboratively. The difference between now and then is that wikinomics has gone beyond a business or a technology trend to a more encompassing societal shift as businesses and communities bypass crumbling institutions and old ways of doing business.

The movement to stop climate change is a good example of mass collaboration in action. We're in the early days of something unprecedented: Thanks to the Web 2.0 the entire world is beginning to collaborate around a single idea for the first time ever—changing the weather. Climate change is quickly becoming a nonpartisan issue, and all citizens obviously have a stake in the outcome. So for the first time we have one global, multimedia, affordable, many-to-many communications system, and one issue on which there is growing consensus. Around the world there are hundreds, probably thousands, of collaborations occurring in which everyone from scientists to schoolchildren are mobilizing to do something about carbon emissions. The "killer application" for mass collaboration may turn out to be saving the planet, literally.

Predictably, however, new paradigms bring dislocation and confusion. They are often received with coolness, or worse—mockery or hostility. Vested interests fight against the shift. Leaders of the old have great difficulty embracing the new.

Wikinomics has certainly had its detractors.

For example, Nicholas Carr, a former editor of the *Harvard Business Review*, wrote an article entitled "The Ignorance of Crowds," in which he criticized us and others for promoting the power of mass collaboration. Carr argues that "peer production is best viewed as a means for refining the old rather than inventing the new; that it's an optimization model more than an invention model."[1] He concludes that only a relatively small and formally organized group of talented professionals can produce breakthroughs.

Essentially the same argument was floated more recently by Jaron Lanier in his book *You Are Not a Gadget*. Unlike some critics, Lanier has a lot of street cred. As a forerunner in virtual reality, he can't be so easily dismissed as a Luddite by technology evangelists, which makes his views worth considering in more detail. Like Carr, Lanier writes that Internet-enabled collaborations produce mediocre outcomes when compared with the secretive, closed-shop approach to innovation that dominated the previous century. "When you have everyone collaborate on everything," Lanier argues, "you generate a dull, average outcome in all things. You don't get innovation." Lanier dismisses Linux, the open source operating system, as "ordinary," and claims that the most sophisticated, influential, and lucrative examples of technology stem from proprietary development.

As this book explains, Lanier is wrong about Linux. When Helsinki-born Linus Torvalds first posted a fledgling version of Linux on an obscure software bulletin board, no one—apart from the most diehard open-source evangelists—would have predicted that open-source software would be much more than a short-lived hackers' experiment. And yet, within a few short years Linux became the largest software engineering project on the planet and spawned a multibillion-dollar ecosystem that upset the balance of power in the software industry.

Today, Linux is used in everything from the smallest consumer electronics to the largest supercomputers. It helps run Germany's air traffic control systems. It also runs a number of nuclear power plants (whose names cannot be disclosed for reasons of national security). If you drive a BMW, chances are it is running Linux. And, at the time of writing, over 500 million users of set-top cable boxes, Tivos, Android phones. and other home appliances use Linux, and over 1.5 billion people use it indirectly every day whenever they access Google, Yahoo, or myriad other Web sites.

Sure, Linux itself is a free, open-source operating system and as such has no employees, stock options, corporate campuses, or free haircuts. But the estimated size of the Linux economy is roughly $50 billion per annum, including Linux-related hardware, consumer electronics, and related services produced by profit-seeking companies around the world. That's up by a factor of 5 since 2006 and it's more than the GDP of some small countries like Costa Rica, Lebanon, and Bolivia.

Linux aside, Lanier and other critics mischaracterize the nature of innovation today and fail to appreciate that mass collaboration is not about "everyone doing everything"; it's about bringing together the complementary skills and knowledge required to create a new product or solve a problem. Take Apple's iPhone, which Lanier erroneously singles out as the epitome of "closed-shop" development. The iPhone is, in fact, the result of a massive networked-based collaboration involving thousands of companies. Various partners help to design the product. A Taiwanese company does the technical specs, manufacturing, and assembly collaborating with hundreds of their own suppliers. And most of the software – supposedly Apple's main source of competitive advantage – is developed not by Apple, but by an army of third-party developers who have created more than 150,000 applications for the App Store.

Other critics call *Wikinomics* a new form of "online collectivism" that threatens our economy and our culture. In *The Cult of the Amateur: How Today's Internet Is Killing Our Culture*, pundit Andrew Keen issues a long bleat claiming that the world is awash with drivel because it is so easy to propagate ideas on the Internet. He laments the fact that newspaper circulation figures are declining while YouTube videos of skateboarding accidents capture the nation's eyeballs. Keen pines for the good old days—a time when apparently only smart and credible people had access to printing presses and the airwaves. A healthy society needs gatekeepers, he argues. Only then can we ensure quality, so, for example, TV commercials will continue being produced by experienced and talented directors and viewers won't be subjected to pathetic wannabe commercials that win contests run by companies looking for cheap advertising.

The irony, of course, is that Keen's critique of the Web 2.0 could just as easily be leveled at the world of professional publishing with its armies of professional editors, fact-checkers, information technology and databases, engineered business processes, reward systems, and all the trappings of an old-style hierarchy. As a recent study by *Nature* revealed, all of our systems of knowledge production have flaws. *Encyclopedia Britannica* was found to have nearly as many errors as Wikipedia—the key difference is that Wikipedia's fluid content creation mechanisms and large volunteer community ensure that its errors get fixed quickly. Amateurs are not drowning out informed and fact-based commentary; they're just replacing the old one-way monologue with increasingly rich and variegated conversations where a billion voices can be heard.

Some critics have argued that *Wikinomics*—with its emphasis on sharing intellectual property—is akin to communism and thus undermines the legitimate right of companies to make a profit. Still others suggest that we are promoting a "free economy" where unpaid volunteers are exploited by corporations. Few of these critics seem to have noticed, however, that the majority of the people who participate in collaborative innovation communities are profiting, sometimes monetarily and other times by using their experience to further their careers or expand their networks. Many Linux programmers now get paid, usually by large companies like IBM and Intel that assign them to help out. YouTube recently began sharing its ad revenue with the contributors of the most popular videos. Scientists who solve problems on InnoCentive, meanwhile, get anything from $10,000 to $1 million for the right solution. In response to the furor over file sharing, artists like Nine Inch Nails and the Beastie Boys have created compelling online venues in which fans can build communities around sharing and remixing their content. In every one of the seven new models of mass collaboration in the book, people and companies are winning.

However, like Lanier and Keen, we are very sympathetic to the need for everyone from artists and writers to scientists and filmmakers to be compensated appropriately for their work. As creators, they should be able to control how their creative works are disseminated and repurposed on the Internet. For some people, that means giving it away to maximize exposure and reap rewards through some complementary avenue. But we tend to agree that the world could end up a poorer place if nobody pays directly for high-quality content anymore.

The turmoil in the media and entertainment industry is evidence of the fact that transitions from one era to the next is never smooth and the long-terms effects are often hard to forecast with accuracy. As media analyst and author Clay Shirky put it in his essay about the impending collapse of the newspaper industry: "That is what real revolutions are like. The old stuff gets broken faster than the new stuff is put in its place." He makes the case that even the revolutionaries can't predict what will happen. "Ancient social bargains, once disrupted, can neither be mended nor quickly replaced, since any such bargain takes decades to solidify," he says.

While that may be true, it is also worth remembering that the future is not something to predict, it is something to achieve.

We readily acknowledge that success with wikinomics will not be easy or automatic. Leaders must rewire their brains to think differently about the business world and resist the temptation to put up the barricades and fight the forces of self-organization in their industries.

With more than a billion individuals online, *Wikinomics* heralds an unprecedented opportunity. Mass collaboration is not only the most significant driver of success in today's marketplaces, it's helping to transform the way we conduct science, create culture, inform and educate ourselves, and govern our communities and nations. Greater openness in innovation and science, for example, is creating more economic opportunity for citizens and businesses that learn how to tap into global innovation webs. In the fight against climate change, new modes of intelligent transport and smart energy grids are introducing new innovation to an outmoded sector and bringing greater consumer awareness and a sense of community to making ordinary household and business decisions that can reduce our carbon footprints. In education, leading universities are building a global network for higher learning that could provide every aspiring student around the globe with access to world-class educational resources that they can return to throughout their lifetime. Pockets of openness in government are generating more productive and equitable services, bolstering public trust and legitimacy, and unlocking new possibilities to co-innovate solutions to local, national and global challenges.

New models of mass collaboration not only underpin these important transformations, they help reveal a new global imperative. Without greater openness and participation in all institutions, the world will be ill equipped to confront the complex challenges that face humanity in the twenty-first century. Harness this new openness, on the other hand, and the world has a better chance to secure prosperity for all its citizens for many more centuries to come. Regardless of the sector,

Organizations that heed the principles of wikinomics are thriving, succeeding, and profiting—not to mention doing good things. We hope you and your organization do so as well.

—Don Tapscott and Anthony Williams
December 2007

CONTENTS

CONTENTS

INTRODUCTION

Throughout history corporations have organized themselves according to strict hierarchical lines of authority. Everyone was a subordinate to someone else—employees versus managers, marketers versus customers, producers versus supply chain subcontractors, companies versus the community. There was always someone or some company in charge, controlling things, at the "top" of the food chain. While hierarchies are not vanishing, profound changes in the nature of technology, demographics, and the global economy are giving rise to powerful new models of production based on community, collaboration, and self-organization rather than on hierarchy and control.

Millions of media buffs now use blogs, wikis, chat rooms, and personal broadcasting to add their voices to a vociferous stream of dialogue and debate called the "blogosphere." Employees drive performance by collaborating with peers across organizational boundaries, creating what we call a "wiki workplace." Customers become "prosumers" by cocreating goods and services rather than simply consuming the end product. So-called supply chains work more effectively when the risk, reward, and capability to complete major projects—including massively complex products like cars, motorcycles, and airplanes—are distributed across planetary networks of partners who work as peers.

Smart companies are encouraging, rather than fighting, the heaving growth of massive online communities—many of which emerged from the fringes of the Web to attract tens of millions of participants overnight. Even ardent competitors are collaborating on path-breaking scientific initiatives that accelerate discovery in their industries. Indeed, as a growing number of firms see the benefits of mass collaboration, this new way of organizing will eventually displace the traditional corporate structures as the economy's primary engine of wealth creation.

Already this new economic model extends beyond software, music, publishing, pharmaceuticals, and other bellwethers to virtually every part of the global economy. But as this process unravels, many managers have concluded that the new mass collaboration is far from benign. Some critics look at successful "open source" projects such as Linux and Wikipedia, for example, and assume they are an attack on the legitimate right and need of companies to make a profit. Others see this new cornucopia of participation in the economy as a threat to their very existence (has anyone bought a music CD lately?).

We paint a very different picture with the evidence we have accumulated in this book. Yes, there are examples of pain and suffering in industries and firms that have so far failed to grasp the new economic logic. But the forthcoming pages are filled with many tales of how ordinary people and firms are linking up in imaginative new ways to drive innovation and success. A number of these stories revolve around the explosive growth of phenomena such as MySpace, InnoCentive, flickr, Second Life, YouTube, and the Human Genome Project. These organizations are harnessing mass collaboration to create real value for participants and have enjoyed phenomenal successes as a result.

Many mature firms are benefiting from this new business paradigm, and we share their stories too. Companies such as Boeing, BMW, and Procter & Gamble have been around for the better part of a century. And yet these organizations and their leaders have seized on collaboration and self-organization as powerful new levers to cut costs, innovate faster, cocreate with customers and partners, and generally do whatever it takes to usher their organizations into the twenty-first-century business environment.

This book, too, is the product of several long-running collaborations. In the last few years the New Paradigm team has conducted several large multiclient investigations to understand how the new Web (sometimes called the Web 2.0) changes the corporation and how companies innovate, build relationships, market, and compete.

A $3 million study in 2000–2001 examined the rise of an increasingly mobile and pervasive Web and its impact on business models.[1] In 2003 we raised $2 million to study Web-enabled transparency as a new force to foster powerful networked businesses and trust.[2] In 2004–2005 a $4 million program explored how new technology and collaborative models change business designs and competitive dynamics.[3]

The conclusion from all of this work is striking and enormously positive. Billions of connected individuals can now actively participate in innovation, wealth creation, and social development in ways we once only dreamed of. And when these masses of people collaborate they collectively can advance the arts, culture, science, education, government, and the economy in surprising but ultimately profitable ways. Companies that engage with these exploding Web-enabled communities are already discovering the true dividends of collective capability and genius.

To succeed, it will not be sufficient to simply intensify existing management strategies. Leaders must think differently about how to compete and be profitable, and embrace a new art and science of collaboration we call wikinomics. This is more than open source, social networking, so-called crowdsourcing, smart mobs, crowd wisdom, or other ideas that touch upon the subject. Rather, we are talking about deep changes in the structure and modus operandi of the corporation and our economy, based on new competitive principles such as openness, peering, sharing, and acting globally.

The results of this foundational research are proprietary to the members that funded it, including more than one hundred in-depth reports and countless executive briefings, seminars, and workshops. However, our work with these companies inspired us to devote weekends and evenings to write a book that would take this work to the next level and inspire a broad audience to apply its ideas, frameworks, and guidelines. We conducted more than one hundred interviews and discussions with key players in this revolution. Unless otherwise footnoted, all quotes in this book come from these conversations.

In the process, we, as authors, learned something about collaboration too. We authored these pages on separate continents, with Don working primarily from Toronto, Canada, and Anthony based in London, England. When we were both working on the manuscript at the same time we hooked up with a Skype connection, talking, exchanging material, or being silent as appropriate. At times it felt like we were in the same room.

We have also collaborated with more than one hundred leading thinkers and practitioners. Their roles in bringing this book to life are graciously acknowledged at the end of the book. In one interesting twist we decided that the best way to come up with a great subtitle was to hold an open discussion on the Web. Within twenty-four hours we had dozens of great subtitle suggestions—the best of which are listed on the Subtitles page.

SUBTITLES

Books have a title page. This is our subtitle page. In what we believe to be a first, we're listing a few of our favorite suggestions for subtitles gleaned from a public online discussion held the week of June 2, 2006. We received more than one hundred great suggestions in the first forty-eight hours. To our collaborators—you know who you are—we extend our most sincere thanks.

Edit This Book!
The Dividends of Collective Genius
We the People
Business (The Remix)
The New World of Collaborative Production
Peer Innovation in the Age of MySpace, Linux, and Wikipedia
Profiting from Collaborative Anarchy
Please Register to Participate
The Power of Us
Creating a New Page in Business History
Unleashing Our Collective Genius
This Book Is a Stub
Harnessing the Power of Your Peers
(Your Input Needed Here)*
Peer-Powered Profit in Life, Business, and Individual Choice
The Peer Advantage: Myth or Magic?
Peer Producing the Future

1. WIKINOMICS

The Art and Science of Peer Production

It was late in the afternoon, on a typically harsh Canadian winter day, as Rob McEwen, the CEO of Goldcorp Inc., stood at the head of the boardroom table confronting a room full of senior geologists. The news he was about to deliver was not good. In fact it was disastrous, and McEwen was having a hard time shielding his frustration.

The small Toronto-based gold-mining firm was struggling, besieged by strikes, lingering debts, and an exceedingly high cost of production, which had caused them to cease mining operations. Conditions in the marketplace were hardly favorable. The gold market was contracting, and most analysts assumed that the company's fifty-year-old mine in Red Lake, Ontario, was dying. Without evidence of substantial new gold deposits, the mine seemed destined for closure, and Goldcorp was likely to go down with it.

Tensions were running at fever pitch. McEwen had no real experience in the extractive industries, let alone in gold mining. Nevertheless, as an adventurous young mutual fund manager he had gotten involved in a takeover battle and emerged as Goldcorp Inc.'s majority owner. Few people in the room had much confidence that McEwen was the right person to rescue the company. But McEwen just shrugged off his critics.

He turned to his geologists and said, "We're going to find more gold on this property, and we won't leave this room tonight until we have a plan to find it." At the conclusion of the meeting he handed his geologists $10 million for further exploration and sent them packing for Northern Ontario.

Most of his staff thought he was crazy but they carried out his instructions, drilling in the deepest and most remote parts of the mine. Amazingly, a few weeks later they arrived back at Goldcorp headquarters beaming with pride and bearing a remarkable discovery: Test drilling suggested rich deposits of new gold, as much as thirty times the amount Goldcorp was currently mining!

The discovery was surprising, and could hardly have been better timed. But after years of further exploration, and to McEwen's deep frustration, the company's geologists struggled to provide an accurate estimate of the gold's value and exact location. He desperately needed to inject the urgency of the market into the glacial processes of an old-economy industry.

In 1999, with the future still uncertain, McEwen took some time out for personal development. He wound up at an MIT conference for young presidents when coincidentally the subject of Linux came up. Perched in the lecture hall, McEwen listened intently to the remarkable story of how Linus Torvalds and a loose volunteer brigade of software developers had assembled the world-class computer operating system over the Internet. The lecturer explained how Torvalds revealed his code to the world, allowing thousands of anonymous programmers to vet it and make contributions of their own.

McEwen had an epiphany and sat back in his chair to contemplate. If Goldcorp employees couldn't find the Red Lake gold, maybe someone else could. And maybe the key to finding those people was to open up the exploration process in the same way Torvalds "open sourced" Linux.

McEwen raced back to Toronto to present the idea to his head geologist. "I'd like to take all of our geology, all the data we have that goes back to 1948, and put it into a file and share it with the world," he said. "Then we'll ask the world to tell us where we're going to find the next six million ounces of gold." McEwen saw this as an opportunity to harness some of the best minds in the industry. Perhaps understandably, the in-house geologists were just a little skeptical.

Mining is an intensely secretive industry, and apart from the minerals themselves, geological data is the most precious and carefully guarded resource. It's like the Cadbury secret—it's just not something companies go around sharing. Goldcorp employees wondered whether the global community of geologists would respond to Goldcorp's call in the same way that software developers rallied around Linus Torvalds. Moreover, they worried about how the contest would reflect on them and their inability to find the illusive gold deposits.

McEwen acknowledges in retrospect that the strategy was controversial and risky. "We were attacking a fundamental assumption; you simply don't give away proprietary data," he said. "It's so fundamental," he adds, "that no one had ever questioned it." Once again, McEwen was determined to soldier on.

In March 2000, the "Goldcorp Challenge" was launched with a total of $575,000 in prize money available to participants with the best methods and estimates. Every scrap of information (some four hundred megabytes worth) about the 55,000–acre property was revealed on Goldcorp's Web site. News of the contest spread quickly around the Internet, as more than one thousand virtual prospectors from fifty countries got busy crunching the data.

Within weeks, submissions from around the world came flooding in to Goldcorp headquarters. As expected, geologists got involved. But entries came from surprising sources, including graduate students, consultants, mathematicians, and military officers, all seeking a piece of the action. "We had applied math, advanced physics, intelligent systems, computer graphics, and organic solutions to inorganic problems. There were capabilities I had never seen before in the industry," says McEwen. "When I saw the computer graphics I almost fell out of my chair." The contestants had identified 110 targets on the Red Lake property, 50 percent of which had not been previously identified by the company. Over 80 percent of the new targets yielded substantial quantities of gold. In fact, since the challenge was initiated an astounding eight million ounces of gold have been found. McEwen estimates the collaborative process shaved two to three years off their exploration time.

Today Goldcorp is reaping the fruits of its open source approach to exploration. Not only did the contest yield copious quantities of gold, it catapulted his underperforming $100 million company into a $9 billion juggernaut while transforming a backward mining site in Northern Ontario into one of the most innovative and profitable properties in the industry. Needless to say McEwen is one happy camper. As are his shareholders. One hundred dollars invested in the company in 1993 is worth over $3,000 today.

Perhaps the most lasting legacy of the Goldcorp Challenge is the validation of an ingenious approach to exploration in what remains a conservative and highly secretive industry. Rob McEwen bucked an industry trend by sharing the company's proprietary data and simultaneously transformed a lumbering exploration process into a modern distributed gold discovery engine that harnessed some of the most talented minds in the field.

McEwen saw things differently. He realized the uniquely qualified minds to make new discoveries were probably outside the boundaries of his organization, and by sharing some intellectual property he could harness the power of collective genius and capability. In doing so he stumbled successfully into the future of innovation, business, and how wealth and just about everything else will be created. Welcome to the new world of wikinomics where collaboration on a mass scale is set to change every institution in society.

THE NEW WORLD OF WIKINOMICS

Due to deep changes in technology, demographics, business, and the world, we are entering a new age where people participate in the economy like never before. This new participation has reached a tipping point where new forms of mass collaboration are changing how goods and services are invented, produced, marketed, and distributed on a global basis. This change presents far-reaching opportunities for every company and for every person who gets connected.

In the past, collaboration was mostly small scale. It was something that took place among relatives, friends, and associates in households, communities, and workplaces. In relatively rare instances, collaboration approached mass scale, but this was mainly in short bursts of political action. Think of the Vietnam-era war protests or, more recently, about the raucous antiglobalization rallies in Seattle, Turin, and Washington. Never before, however, have individuals had the power or opportunity to link up in loose networks of peers to produce goods and services in a very tangible and ongoing way.

Most people were confined to relatively limited economic roles, whether as passive consumers of mass-produced products or employees trapped deep within organizational bureaucracies where the boss told them what to do. Even their elected representatives barely concealed their contempt for bottom-up participation in decision making. In all, too many people were bypassed in the circulation of knowledge, power, and capital, and thus participated at the economy's margins.

Today the tables are turning. The growing accessibility of information technologies puts the tools required to collaborate, create value, and compete at everybody's fingertips. This liberates people to participate in innovation and wealth creation within every sector of the economy. Millions of people already join forces in self-organized collaborations that produce dynamic new goods and services that rival those of the world's largest and best-financed enterprises. This new mode of innovation and value creation is called "peer production," or peering—which describes what happens when masses of people and firms collaborate openly to drive innovation and growth in their industries.[1]

Some examples of peer production have recently become household names. As of August 2006, the online networking extravaganza MySpace had one hundred million users—growing at half a million a week—whose personal musings, connections, and profiles are the primary engines of value creation on the site. MySpace, YouTube, Linux, and Wikipedia—today's exemplars of mass collaboration—are just the beginning; a few familiar characters in the opening pages of the first chapter in a long-running saga that will change many aspects of how the economy operates. In the forthcoming pages of this book we describe seven unique forms of peer production that are making the economy more dynamic and productive. Along the way we offer engaging stories for the casual reader and great insights for the businessperson seeking to harness this new force in their business.

Age of Participation

Call them the "weapons of mass collaboration." New low-cost collaborative infrastructures—from free Internet telephony to open source software to global outsourcing platforms—allow thousands upon thousands of individuals and small producers to cocreate products, access markets, and delight customers in ways that only large corporations could manage in the past. This is giving rise to new collaborative capabilities and business models that will empower the prepared firm and destroy those that fail to adjust.

The upheaval occurring right now in media and entertainment provides an early example of how mass collaboration is turning the economy upside down. Once a bastion of "professionalism," credentialed knowledge producers share the stage with "amateur" creators who are disrupting every activity they touch. Tens of millions of people share their news, information, and views in the blogosphere, a self-organized network of over one hundred million personal commentary sites that are updated every second of the day.[2] Some of the largest weblogs (or blogs for short) receive a half a million daily visitors,[3] rivaling some daily newspapers. Now audioblogs, podcasts, and mobile photo blogs are adding to a dynamic, up-to-the-minute stream of person-to-person news and information delivered free over the Web.

Individuals now share knowledge, computing power, bandwidth, and other resources to create a wide array of free and open source goods and services that anyone can use or modify. What's more, people can contribute to the "digital commons" at very little cost to themselves, which makes collective action much more attractive. Indeed, peer production is a very social activity. All one needs is a computer, a network connection, and a bright spark of initiative and creativity to join in the economy.

These new collaborations will not only serve commercial interests, they will help people do public-spirited things like cure genetic diseases, predict global climate change, and find new planets and stars. Researchers at Olson Laboratory, for example, use a massive supercomputer to evaluate drug candidates that might one day cure AIDS. This is no ordinary supercomputer, however. Their FightAIDS@home initiative is part of the World Community Grid, a global network where millions of individual computer users donate their spare computing power via the Internet to form one of the world's most powerful computing platforms.

These changes, among others, are ushering us toward a world where knowledge, power, and productive capability will be more dispersed than at any time in our history—a world where value creation will be fast, fluid, and persistently disruptive. A world where only the connected will survive. A power shift is underway, and a tough new business rule is emerging: Harness the new collaboration or perish. Those who fail to grasp this will find themselves ever more isolated—cut off from the networks that are sharing, adapting, and updating knowledge to create value.

This might sound like hyperbole, but it's not. Consider some additional ways ordinary citizens can now participate in the global body économique.

Rather than just read a book, you can write one. Just log on to Wikipedia—a collaboratively created encyclopedia, owned by no one and authored by tens of thousands of enthusiasts. With five full-time employees, it is ten times bigger than *Encyclopedia Britannica* and roughly the same in accuracy.[4] It runs on a wiki, software that enables users to edit the content of Web pages. Despite the risks inherent in an open encyclopedia in which everyone can add their views, and constant battles with detractors and saboteurs, Wikipedia continues to grow rapidly in scope, quality, and traffic. The English-language version has more than three million entries, and there are over two hundred sister sites in languages ranging from Polish and Japanese to Hebrew and Catalan.

Or perhaps your thing is chemistry. Indeed, if you're a retired, unemployed, or aspiring chemist, Procter & Gamble needs your help. The pace of innovation has doubled in its industry in the past five years alone, and now its army of 7,500 researchers is no longer enough to sustain its lead. Rather than hire more researchers, CEO A. G. Lafley instructed business unit leaders to source 50 percent of their new product and service ideas from outside the company. Now you can work for P&G without being on their payroll. Just register on the InnoCentive network where you and two hundred thousand other scientists around the world can help solve tough R&D problems for a cash reward. InnoCentive is only one of many revolutionary marketplaces matching scientists to R&D challenges presented by companies in search of innovation. P&G and thousands of other companies look to these marketplaces for ideas, inventions, and uniquely qualified minds that can unlock new value in their markets.

Media buffs are similarly empowered. Rather than consume the TV news, you can now create it, along with thousands of independent citizen journalists who are turning the profession upside down. Tired of the familiar old faces and blather on network news? Turn off your TV, pick up a video camera and some cheap editing software, and create a news feature for Current TV, a new national cable and satellite network created almost entirely by amateur contributors. Though the contributors are unpaid volunteers, the content is surprisingly good. Current TV provides online tutorials for camera operation and storytelling techniques, and their guidelines for creating stories help get participants started. Viewers vote on which stories go to air, so only the most engaging material makes prime time.

Finally, a young person in India, China, Brazil, or any one of a number of emerging Eastern European countries can now do what their parents only dreamed of by joining the global economy on an equal footing. You might be in a call center in Bangalore that takes food orders for a drive-through restaurant in Los Angeles. Or you could find yourself working in Foxconn's new corporate city in the Schenzen province of China, where a decade ago farmers tilled the land with oxen. Today 180,000 people work, live, learn, and play on Foxconn's massive high-tech campus, designing and building consumer electronics for teenagers around the globe.

For incumbents in every industry this new cornucopia of participation and collaboration is both exhilarating and alarming. As New Paradigm executive David Ticoll argues, "Not all examples of self-organization are benign, or exploitable. Within a single industry the development of opportunities for self-organized collaboration can be beneficial, neutral, or highly competitive to individual firms, or some combination of at least two of these." Publishers found this out the hard way. Blogs, wikis, chat rooms, search engines, advertising auctions, peer-to-peer downloading, and personal broadcasting represent new ways to entertain, communicate, and transact. In each instance the traditionally passive buyers of editorial and advertising take active, participatory roles in value creation. Some of these grassroots innovations pose dire threats to existing business models.

Publishers of music, literature, movies, software, and television are like proverbial canaries in a coal mine—the first casualties of a revolution that is sweeping across all industries. Many enfeebled titans of the industrial economy feel threatened. Despite heroic efforts to change, they remain shackled by command-and-control legacies. Companies have spent the last three decades remolding their operations to compete in a hyper-competitive economy—ripping costs out of their businesses at every opportunity; trying to become more "customer-friendly"; assembling global production networks; and scattering their bricks-and-mortar R&D organizations around the world.

Now, to great chagrin, industrial-era titans are learning that the real revolution is just getting started. Except this time the competition is no longer their arch industry rivals; it's the uberconnected, amorphous mass of self-organized individuals that is gripping their economic needs firmly in one hand, and their economic destinies in the other. "We the People" is no longer just a political expression—a hopeful ode to the power of "the masses"; it's also an apt description of how ordinary people, as employees, consumers, community members, and taxpayers now have the power to innovate and to create value on the global stage.

For smart companies, the rising tide of mass collaboration offers vast opportunity. As the Goldcorp story denotes, even the oldest of old economy industries can harness this revolution to create value in unconventional ways. Companies can reach beyond their walls to sow the seeds of innovation and harvest a bountiful crop. Indeed, firms that cultivate nimble, trust-based relationships with external collaborators are positioned to form vibrant business ecosystems that create value more effectively than hierarchically organized businesses.

For individuals and small producers, this may be the birth of a new era, perhaps even a golden one, on par with the Italian renaissance or the rise of Athenian democracy. Mass collaboration across borders, disciplines, and cultures is at once economical and enjoyable. We can peer produce an operating system, an encyclopedia, the media, a mutual fund, and even physical things like a motorcycle. We are becoming an economy unto ourselves—a vast global network of specialized producers that swap and exchange services for entertainment, sustenance, and learning. A new economic democracy is emerging in which we all have a lead role.

Promise and Peril

Experience shows that the first wave of Internet-enabled change was tainted by irrational exuberance. A sober analysis of today's trends reveals that this new participation is both a blessing and a curse. Mass collaboration can empower a growing cohort of connected individuals and organizations to create extraordinary wealth and reach unprecedented heights in learning and scientific discovery. If we are wise, we will harness this capability to create opportunities for everyone and to carefully steward the planet's natural resources. But the new participation will also cause great upheaval, dislocation, and danger for societies, corporations, and individuals that fail to keep up with relentless change.

As with all previous economic revolutions, the demands on individuals, organizations, and nations will be intense, and at times traumatic, as old industries and ways of life give way to new processes, technologies, and business models. The playing field has been ripped wide open, and the recurrent need to reconfigure people and capabilities to serve an ever-changing market will require individuals to embrace constant change and renewal in their careers.

As recent events foretell, a smaller, more open and interdependent world has the potential to be dynamic and vibrant, but also more vulnerable to terrorism and criminal networks. Just as the masses of scientists and software coders can collaborate on socially beneficial projects, criminals and terrorists can conspire over the Internet to wreak havoc on our daily existence.

Even with good intentions, mass collaboration is certainly no panacea. When people organize en masse to create goods, services, and entertainment they create new challenges as well as opportunities. Renowned computer scientist, composer, and author Jaron Lanier worries that collaborative communities such as flickr, MySpace, and Wikipedia represent a new form of "online collectivism" that is suffocating authentic voices in a muddled and anonymous tide of mass mediocrity. Lanier laments the idea that "the collective is all-wise," or as he put it, "that it is desirable to have influence concentrated in a bottleneck that can channel the collective with the most verity and force." He rightly points out that such ideas have had terrible consequences when imposed by ruthless dictators like Stalin or Pol Pot. But his argument runs afoul when he attributes the same kind of "collective stupidity" to the emerging forms of mass collaboration on the Web.

Other wise and thoughtful people such as Microsoft's Bill Gates, meanwhile, complain that the incentives for knowledge producers are disappearing in a world where individuals can pool their talents to create free goods that compete with proprietary marketplace offerings. Gates cites the movement to assemble a global "creative commons" that contains large bodies of scientific and cultural content as a potential threat to the ability to make profits in knowledge-based industries such as software. Many top executives are lining up alongside Gates to harpoon what they see as new-fangled "communists" in various guises.

Reactionary sentiments are hardly surprising given the circumstances. The production of knowledge, goods, and services is becoming a collaborative activity in which growing numbers of people can participate. This threatens to displace entrenched interests that have prospered under the protection of various barriers to entry, including the high costs of obtaining the financial, physical, and human capital necessary to compete. Companies accustomed to comfortably directing marketplace activities must contend with new and unfamiliar sources of competition, including the self-organized masses, just as people in elite positions (whether journalists, professors, pundits, or politicians) must now work harder to justify their exalted status. As the global division of labor becomes ever more complex, variegated, and dynamic, the economy is spinning out of the control of the usual suspects. There will be casualties, but the winners will outnumber the losers. Indeed, we believe the new era heralds more economic opportunity for individuals and businesses, and greater efficiency, creativity, and innovation throughout the economy as a whole.

Though we disagree with Lanier and Gates, they do raise important issues that we will address throughout this book. For now it must be said that mass collaboration and peer production are really the polar opposites of the communism that Gates and Lanier despise. Digital pioneer Howard Rheingold points out, "Collectivism involves coercion and centralized control; collective action involves freely chosen self-selection and distributed coordination." Whereas communism stifled individualism, mass collaboration is based on individuals and companies employing widely distributed computation and communication technologies to achieve shared outcomes through loose voluntary associations.

What's more, the participation revolution now underway opens up new possibilities for billions of people to play active roles in their workplaces, communities, national democracies, and the global economy at large. This has profound social benefits, including the opportunity to make governments more accountable and lift millions of people out of poverty.

Moreover, it is wrong to assume that the new collective action represents only a threat to established businesses. While some fear mass collaboration will reduce the proportion of our economy that is available for profitable activity and wealth creation, we will show the opposite. New models of peer production can bring the prepared manager rich new possibilities to unlock innovative potential in a wide range of resources that thrive inside and outside the firm. With the right approach, companies can obtain higher rates of growth and innovation by learning how to engage and cocreate with a dynamic and increasingly global network of peers. Rather than conceding defeat to the most powerful economic force of our times, established companies can harness the new collaboration for unparalleled success.

A new art and science of collaboration is emerging—we call it "wikinomics." We're not just talking about creating online encyclopedias and other documents. A wiki is more than just software for enabling multiple people to edit Web sites. It is a metaphor for a new era of collaboration and participation, one that, as Dylan sings, "will soon shake your windows and rattle your walls." The times are, in fact, a changin'.

THE NEW PROMISE OF COLLABORATION

Word association test: What's the first thing that comes to mind when you hear the word "collaboration"? If you're like most people, you conjure up images of people working together happily and productively. In everyday life, we collaborate with fellow parents at a PTA meeting, with other students on a class project, or with neighbors to protect and enhance our communities. In business we collaborate with coworkers at the office, with partners in the supply chain, and within teams that traverse departmental and organizational silos. We collaborate on research projects, work together to make a big sale, or plan a new marketing campaign.

Google CEO Eric Schmidt says, "When you say 'collaboration,' the average forty-five-year-old thinks they know what you're talking about—teams sitting down, having a nice conversation with nice objectives and a nice attitude. That's what collaboration means to most people."

We're talking about something dramatically different. The new promise of collaboration is that with peer production we will harness human skill, ingenuity, and intelligence more efficiently and effectively than anything we have witnessed previously. Sounds like a tall order. But the collective knowledge, capability, and resources embodied within broad horizontal networks of participants can be mobilized to accomplish much more than one firm acting alone. Whether designing an airplane, assembling a motorcycle, or analyzing the human genome, the ability to integrate the talents of dispersed individuals and organizations is becoming *the* defining competency for managers and firms. And in the years to come, this new mode of peer production will displace traditional corporation hierarchies as the key engine of wealth creation in the economy.

In Chapter 2, we discuss a variety of social, economic, and demographic forces that are fueling the rising tide of mass collaboration. More than anything, however, the evolution of the Internet is driving this new age. From the stunning increases in computing power, network capability, and reach, to the growing accessibility of the tools required to get organized, create value, and compete, this new Web has opened the floodgates to a worldwide explosion of participation.

There are many names for this new Web: the Web 2.0, the living Web, the Hypernet, the active Web, the read/write Web.[5] Call it what you like—the sentiment is the same. We're all participating in the rise of a global, ubiquitous platform for computation and collaboration that is reshaping nearly every aspect of human affairs. While the old Web was about Web sites, clicks, and "eyeballs," the new Web is about the communities, participation, and peering. As users and computing power multiply, and easy-to-use tools proliferate, the Internet is evolving into a global, living, networked computer that anyone can program. Even the simple act of participating in an online community makes a contribution to the new digital commons—whether one's building a business on Amazon or producing a video clip for YouTube, creating a community around his or her flickr photo collection or editing the astronomy entry on Wikipedia.

This new Web already links more than a billion people directly and (unlike Web 1.0) is reaching out to the physical world, connecting countless inert objects, from hotel doors to cars. It is beginning to deliver dynamic new services—from free long-distance video telephony to remote brain surgery. And it covers the planet like a skin, linking a machine soldering chips onto circuit boards in Singapore with a chip warehouse in Denver, Colorado.

Twenty years from now we will look back at this period of the early twenty-first century as a critical turning point in economic and social history. We will understand that we entered a new age, one based on new principles, worldviews, and business models where the nature of the game was changed.

The pace of change and the evolving demands of customers are such that firms can no longer depend only on internal capabilities to meet external needs. Nor can they depend only on tightly coupled relationships with a handful of business partners to keep up with customer desires for speed, innovation, and control. Instead, firms must engage and cocreate in a dynamic fashion with everyone—partners, competitors, educators, government, and, most of all, customers.

To innovate and succeed, the new mass collaboration must become part of every leader's playbook and lexicon. Learning how to engage and cocreate with a shifting set of self-organized partners is becoming an essential skill, as important as budgeting, R&D, and planning.

THE PRINCIPLES OF WIKINOMICS

The new mass collaboration is changing how companies and societies harness knowledge and capability to innovate and create value. This affects just about every sector of society and every aspect of management. A new kind of business is emerging—one that opens its doors to the world, coinnovates with everyone (especially customers), shares resources that were previously closely guarded, harnesses the power of mass collaboration, and behaves not as a multinational but as something new: a truly global firm. These companies are driving important changes in their industries and rewriting many rules of competition.

Now compare this to traditional business thinking. Conventional wisdom says companies innovate, differentiate, and compete by doing certain things right: by having superior human capital; protecting their intellectual property fiercely; focusing on customers; thinking globally but acting locally; and by executing well (i.e., having good management and controls). But the new business world is rendering each of these principles insufficient, and in some cases, completely inappropriate.

The new art and science of wikinomics is based on four powerful new ideas: openness, peering, sharing, and acting globally. These new principles are replacing some of the old tenets of business. Our objective throughout this book is to provide vivid examples of how people and organizations are harnessing these principles to drive innovation in their workplaces, communities, and industries.

Being Open

If you consider the vernacular, the term "open" is loaded—rich with meaning and positive connotations. Among other things, openness is associated with candor, transparency, freedom, flexibility, expansiveness, engagement, and access. Open, however, is not an adjective often used to describe the traditional firm, and until recently, open would not have appropriately described the inner workings of the economy either. Recently, smart companies have been rethinking openness, and this is beginning to affect a number of important functions, including human resources, innovation, industry standards, and communications.

Companies were closed in their attitudes toward networking, sharing, and encouraging self-organization, in large part because conventional wisdom says that companies compete by holding their most coveted resources close to their chest. When it came to managing human resources, firms were exhorted to hire the best people, and to motivate, develop, and retain them, since human capital is the foundation of competitiveness. Today companies that make their boundaries porous to external ideas and human capital outperform companies that rely solely on their internal resources and capabilities.

Rapid scientific and technological advances are among the key reasons why this new openness is surfacing as a new imperative for managers. Most businesses can barely manage to research the fundamental disciplines that contribute to their products, let alone retain the field's most talented people within their boundaries. So to ensure they remain at the forefront of their industries, companies must increasingly open their doors to the global talent pool that thrives outside their walls.[6]

Standards are another area where openness is gaining momentum. In today's complex and fast-moving economy, the economic deficiencies and liabilities caused by the lack of standardization surface faster, and they are more jarring and consequential than in the past. For years the information technology (IT) industry fiercely fought concepts like open systems and open source. But in the last decade there has been a stampede toward open standards, in part because customers are demanding them. Customers were fed up with being locked into each vendor's architecture, where applications were islands and not portable to another vendor's hardware. Microsoft reaped huge revenues as the provider of a standard platform on which software companies could build their applications, regardless of the brand name on the computer. The shift to openness gained momentum as IT professionals began to collaborate on a wide range of open software platforms. The result was Apache for Web servers, Linux for operating systems, MySQL for databases, Firefox for browsers, and the World Wide Web itself.

Yet another kind of openness is exploding: the communication of previously secret corporate information to partners, employees, customers and shareholders, and other interested participants. Transparency—the disclosure of pertinent information—is a growing force in the networked economy. This goes far beyond the obligation to comply with laws regarding the disclosure of financial information. This is not about the Securities and Exchange Commission (SEC), Sarbanes-Oxley, Eliot Spitzer, or avoiding the "perp walk." Rather, people and institutions that interact with firms are gaining unprecedented access to important information about corporate behavior, operations, and performance. Armed with new tools to find out, inform others, and self-organize, stakeholders are scrutinizing the firm like never before.

Customers can see the true value of products better. Employees have previously unthinkable knowledge about their firm's strategy, management, and challenges. Partners must have intimate knowledge about each other's operations to collaborate. Powerful institutional investors who now own or manage most wealth are developing x-ray vision. And in a world of instant communications, whistle-blowers, inquisitive media, and Googling, citizens and communities can easily put firms under the microscope.

Leading firms are opening up pertinent information to all these groups—because they reap significant benefits from doing so. Rather than something to be feared, transparency is a powerful new force for business success. Smart firms embrace transparency and are actively open. Our research shows that transparency is critical to business partnerships, lowering transaction costs between firms and speeding up the metabolism of business webs. Employees of open enterprises have higher trust among each other and with the firm, resulting in lower costs, better innovation, and loyalty. And when companies like Progressive Insurance are open with customers—honestly sharing both their prices and their competitors', even when they are not as good—customers respond by giving their trust.[7]

Finally, it's worth noting that the economy and society are open in new ways too. Falling trade barriers and information technologies are often cited as key reasons why dozens of highly competitive countries have entered the global economy for the first time, but take education as another important example. Today an aspiring student in Mumbai who has always dreamed of going to MIT can now access the university's entire curriculum online without paying a penny in tuition fees. She can just log on to ocw.mit.edu, and she will read "Welcome to MIT's OpenCourseWare: a free and open educational resource (OER) for educators, students, and self-learners around the world. MIT OpenCourseWare (MIT OCW) supports MIT's mission to advance knowledge and education, and serve the world in the 21st century." She can engage with the content and faculty of one of the world's leading universities, studying everything from aeronautics to zoology. Download the readings and assignments for courses. Share her experiences in one of the community forums. Become part of MIT, participating in lifelong learning for the global knowledge economy.

Peering

Throughout most of human history, hierarchies of one form or another have served as the primary engines of wealth creation and provided a model for institutions such as the church, the military, and government. So pervasive and enduring has the hierarchical mode of organization been that most people assume that there are no viable alternatives. Whether the ancient slave empires of Greece, Rome, China, and the Americas, the feudal kingdoms that later covered the planet, or the capitalist corporation, hierarchies have organized people into layers of superiors and subordinates to fulfill both public and private objectives. Even the management literature today that advocates empowerment, teams, and enlightened management techniques takes as a basic premise the command modus operandi inherent in the modern corporation. Though it is unlikely that hierarchies will disappear in the foreseeable future, a new form of horizontal organization is emerging that rivals the hierarchical firm in its capacity to create information-based products and services, and in some cases, physical things. As mentioned, this new form of organization is known as peering.

The quintessential example of peering is Linux, which we introduced briefly during the Goldcorp story. While the basic facts of Linux are well known in the technology community, they are not known by all, so allow us to briefly recap the story. In 1991, before the World Wide Web had even been invented, a young programmer from Helsinki named Linus

Torvalds created a simple version of the Unix operating system. He called it Linux and shared it with other programmers via an online bulletin board. Of the first ten programmers who corresponded with him, five made substantive changes. Torvalds eventually decided to license the operating system under a general public license (GPL) so that anyone could use it for free, provided they made their changes to the program available to others. Over time an informal organization emerged to manage ongoing development of the software that continues to harness inputs from thousands of volunteer programmers. Because it was reliable and free, Linux became a useful operating system for computers hosting Web servers, and ultimately databases, and today many companies consider Linux an enterprise software keystone.

Today the growing ease with which people can collaborate opens up the economy to new Linux-like projects everyday. People increasingly self-organize to design goods or services, create knowledge, or simply produce dynamic, shared experiences. A growing number of examples suggest that peer-to-peer models of organizing economic activity are making inroads into areas that go well beyond creating software. Take two examples for starters.

Researchers at the Australian biotech institute CAMBIA worry that patents owned by multinational firms such as Monsanto are compromising billions of people who can't afford the licensing fees to exploit genetically modified crops. So CAMBIA researchers who are working on solutions to the challenges of food security and agricultural productivity release their results publicly under BiOS (Biological Open Source Licenses). This way they engage a much wider pool of talented scientists in the process of getting solutions to farmers who need them.

Marketocracy employs a similar form of peering in a mutual fund that harnesses the collective intelligence of the investment community. It has recruited eighty-five thousand traders to manage virtual stock portfolios in a competition to become the best investors. Marketocracy indexes the top one hundred performers, and their trading strategies are emulated in a mutual fund that consistently outperforms the S&P 500. Though not strictly open source, it is an example of how meritocratic, peer-to-peer models are seeping into an industry where conventional wisdom favors the lone superstar stock adviser.

These cases are tangible examples of a new mode of production that is emerging in the heart of the most advanced economies in the world— producing a rich new economic landscape and challenging our basic assumption about human motivation and behavior. In some cases, self-organized "nonmarket" production is moving into arenas that used to be dominated by profit-making firms. Wikipedia, with its free online encyclopedia, is one example where a once vibrant publishing industry is suffering. At the same time, powerful new economic ecosystems are forming on top of shared infrastructures and resources like Linux. Though Linux is free to use or modify, it has been embedded in all kinds of profitable products and services developed by large companies like BMW, IBM, Motorola, Philips, and Sony.

Participants in peer production communities have many different motivations for jumping in, from fun and altruism to achieving something that is of direct value to them. Though egalitarianism is the general rule, most peer networks have an underlying structure, where some people have more authority and influence than others. But the basic rules of operation are about as different from a corporate command-and-control hierarchy as the latter was from the feudal craft shop of the preindustrial economy.

Peering succeeds because it leverages self-organization—a style of production that works more effectively than hierarchical management for certain tasks. Its greatest impact today is in the production of information goods—and its initial effects are most visible in the production of software, media, entertainment, and culture—but there are few reasons for peer production to stop there. Why not open source government? Could we make better decisions if we were to tap the insights of a broader and more representative body of participants? Or perhaps we could apply peer production to physical objects like cars, airplanes, and motorcycles. As we will discover in later chapters, these are not idle fantasies, but real opportunities that the new world of wikinomics makes possible.

Sharing

Conventional wisdom says you should control and protect proprietary resources and innovations—especially intellectual property—through patents, copyright, and trademarks. If someone infringes your IP, get the lawyers out to do battle. Many industries still think this way. Millions of technology-literate kids and teenagers use the Internet to freely create and share MP3 software tools and music. Digital music presents a huge opportunity to place artists and consumers at the center of a vast web of value creation. But rather than embrace MP3 and adopt new business models, the industry has adopted a defensive posture. Obsession with control, piracy, and proprietary standards on the part of large industry players has only served to further alienate and anger music listeners.

No doubt digitization introduces tough new appropriation problems for the creators of digital content. Digital inventions are easy to share, remix, and repurpose, and just as easy to replicate. On the plus side, this means industries with zero marginal cost (i.e., software and digital entertainment) can gain incredible economies of scale. But if your invention can be replicated at no cost, why should anyone pay? And if no one pays, how do you recoup your fixed-cost investment?

Hollywood's proposed solution is to expand the scope and vigor of IP protection. New digital rights management technologies make knowledge and content more excludable—information can be metered, consumer behavior can be controlled, and owners of intellectual property can extract a fee for access. Walled gardens of content, proprietary databases, closed-source software: They all promise healthy returns for knowledge producers. But at the same time, they all restrict access to the essential tools of a knowledge-based economy. And worse, they shut out the real opportunities for customer-driven innovation and creativity that could spawn new business models and industries.

Today, a new economics of intellectual property is prevailing. Increasingly, and to a degree paradoxically, firms in electronics, biotechnology, and other fields find that maintaining and defending a proprietary system of intellectual property often cripples their ability to create value. Smart firms are treating intellectual property like a mutual fund—they manage a balanced portfolio of IP assets, some protected and some shared.

For example, starting in 1999, more than a dozen pharmaceutical firms—hardly what one would call modern-day communists—abandoned their proprietary R&D projects to support open collaborations such as the SNP Consortium and the Alliance for Cellular Signaling (see Chapter 6, The New Alexandrians). Both projects aggregate genetic information culled from biomedical research in publicly accessible databases. They also use their shared infrastructures to harness resources and insights from the for-profit and not-for-profit research worlds. These efforts are speeding the industry toward fundamental breakthroughs in molecular biology— breakthroughs that promise an era of personalized medicine and treatments for intractable disorders. Nobody gives up their potential patent rights over new end products, and by sharing some basic intellectual property the companies bring products to market more quickly.

This logic of sharing applies in virtually every industry. "Just as it's true that a rising tide lifts all boats," says Tim Bray, director of Web technologies at Sun Microsystems, "we genuinely believe that radical sharing is a win-win for everyone. Expanding markets create new opportunities." Under the right conditions, the same could be said of most industries, whether automobiles or other consumer products.

Of course companies need to protect critical intellectual property. They should always protect their crown jewels, for example. But companies can't collaborate effectively if all of their IP is hidden. Contributing to the commons is not altruism; it's often the best way to build vibrant business ecosystems that harness a shared foundation of technology and knowledge to accelerate growth and innovation.

The power of sharing is not limited to intellectual property. It extends to other resources such as computing power, bandwidth, content, and scientific knowledge. Peer-to-peer sharing of computing power, for example, is bringing the telecommunications business to its knees. The cofounder and CEO of Skype, Niklas Zennstrom, says, "The idea of charging for telephone calls belongs to the last century." His company's software harnesses the collective computing power of peers, allowing them to speak with each other free of charge via the Internet. The result is a self-sustaining phone system that requires no central capital investment—just the willingness of its users to share.

The Luxemberg-based Skype went from one hundred thousand to one hundred million registered users in two years, and was acquired by eBay for $2.6 billion in September 2005. The first time Michael Powell, then chairman of the Federal Communications Commission, used Skype, he concluded: "It's over. The world will change now inevitably."

Acting Globally

Consider life on the Galápagos Islands. Its separation from the rest of the world has resulted in a diverse collection of species, many found nowhere else on earth, yet each uniquely tailored to its environment. Now imagine what would happen if a teleportation device appeared on the Galápagos, thereby enabling resident animals to intermingle and roam freely among the islands. Surely the Galápagos would never be the same.

This thought experiment illustrates the consequences of the new era of globalization. The barriers between the Galápagos and the mainland are analogous to geographic and economic barriers that insulate firms and nations. When the insulation is removed, it cannot help but produce disruptive effects on business strategy, enterprise structures, the competitive landscape, and the global social and political order.

Thomas Friedman's book *The World Is Flat* brought the significance of the new globalization to many. But the quickening pace and deep consequences of globalization for innovation and wealth creation are not yet fully understood. In the last twenty years of globalization we have seen Chinese and Indian economic liberalization, the collapse of the Soviet Union, and the first stage of the worldwide information technology revolution. The next twenty years of globalization will help sustain world economic growth, raise world living standards, and substantially deepen global interdependence. At the same time, it will profoundly shake up the status quo almost everywhere—generating enormous economic, cultural, and political convulsions.[8]

On the economic front, the ongoing integration of national economies into a borderless world and the surprisingly fast and furious rise of new titans such as China, India, and South Korea will continue to broaden and flatten the playing field. Two billion more people from Asia and Eastern Europe are already joining the global workforce. And while developed countries worry about growing dependency ratios, most of the increase in world population and consumer demand will take place in today's developing nations—especially in China, India, and Indonesia.

The new globalization is both causing and caused by changes in collaboration and the way firms orchestrate capability to innovate and produce things. Staying globally competitive means monitoring business developments internationally and tapping a much larger global talent pool. Global alliances, human capital marketplaces, and peer production communities will provide access to new markets, ideas, and technologies. People and intellectual assets will need to be managed across cultures, disciplines, and organizational boundaries. Winning companies will need to know the world, including its markets, technologies, and peoples. Those that don't will find themselves handicapped, unable to compete in a business world that is unrecognizable by today's standards.

To do all this, it makes sense to not only *think* globally, as the mantra says, but to *act* globally as well. Managers in the trenches are finding out that acting globally is a tremendous operational challenge, especially when you're buried in legacy systems and processes. Ralph Szygenda, CIO of General Motors, says, "Most big companies are multinationals, not global, and increasingly that's a big problem for all of us."

Szygenda describes how GM grew up as a collection of separate companies. Each major brand, including Cadillac, Oldsmobile, and Buick, had separate staff, procedures, and agendas, and there was very little coordination among them. They might have found shelter under the same umbrella, but they were about as friendly as a group of strangers standing on a New York City sidewalk.

Like many multinationals, GM also was divided into geographically demarcated fiefdoms. Regional divisions had power and autonomy to develop, manufacture, and distribute cars according to local needs and by sourcing from local suppliers. For GM as a whole this federated structure came with immense and costly redundancies, as each division employed a full roster of local workers to take care of everything from manufacturing to human resources. Bob Lutz, GM's vice chairman of global product development, says that duplication of effort cost the company billions of dollars a year and prevented it from leveraging its size and scale.

In an increasingly global and competitive economy such redundancies are swiftly punished. So it pays to have global capabilities—including truly global workforces, unified global processes, and a global IT platform to enhance collaboration among all of the parts of the business as well as the company's web of external partners.

By definition, a truly global company has no physical or regional boundaries. It builds planetary ecosystems for designing, sourcing, asembling, and distributing products on a global basis. The emergence of open IT standards makes it considerably easier to build a global business by integrating best-of-breed components from various geographies.[9] Szygenda envisions how such unity might play out for GM. "Whether we're developing a product, manufacturing, sourcing, or distributing," he says, "we'll be able to link up all of our activities in a seamless global operation." Or as Bob Lutz says, "My vision would be a corporation operating on a truly global basis—no U.S. dominance. We will have global budgets that will be administered optimally, be it the allocation of capital, the allocation of design resources, engineering resources, purchasing, manufacturing. We will treat the whole world as if it were one country."

If companies can go global, how about individuals? In fact, it turns out they can. When we went to see Steve Mills, who runs IBM's software operation, he was immersed in twenty different instant messaging sessions with clients and colleagues around the world. He says, "When computers run fast enough, and the bandwidth is there, everything that is remote feels local—in fact, the whole world feels local to me. I don't need to be present in the room to participate." The new global platform for collaboration opens up myriad new possibilities for individuals like Mills to act globally. The world is teeming with possibilities for education, work, and entrepreneurship—one just needs the skills, motivation, the capacity for lifelong learning, and a basic income level to get connected.

THRIVING IN A WORLD OF WIKINOMICS

These four principles—openness, peering, sharing, and acting globally—increasingly define how twenty-first-century corporations compete. This is very different from the hierarchical, closed, secretive, and insular multinationals that dominated the previous century.

One thing that has not changed is that winning organizations (and societies) will be those that tap the torrent of human knowledge and translate it into new and useful applications. The difference today is that the organizational values, skills, tools, processes, and architectures of the ebbing command-and-control economy are not simply outdated; they are handicaps on the value creation process. In an age where mass collaboration can reshape an industry overnight, the old hierarchical ways of organizing work and innovation do not afford the level of agility, creativity, and connectivity that companies require to remain competitive in today's environment. Every individual now has a role to play in the economy, and every company has a choice—commoditize or get connected.

Changes of this magnitude have occurred before. In fact, human societies have always been punctuated by periods of great change that not only cause people to think and behave differently, but also give rise to new social orders and institutions. In many instances these changes are driven by disruptive technologies, such as the printing press, the automobile, and the telephone, that penetrate societies to fundamentally change their culture and economy.

The new Web—which is really an internetworked constellation of disruptive technologies—is the most robust platform yet for facilitating and accelerating new creative disruptions. People, knowledge, objects, devices, and intelligent agents are converging in many-to-many networks where new innovations and social trends spread with viral intensity. Organizations that have scrambled to come up with responses to new phenomena like Napster or the blogosphere should expect much more of the same—at an increasing rate—in the future.

Previous technology-driven revolutions, like the electrification of industry, took the better part of a century to unfold.[10] Today the escalating scope and scale of the resources applied to innovation means that change will unfold more quickly. Though we are still just beginning a profound economic and institutional adjustment, incumbents should not expect a grace period. The old, hardwired "plan and push" mentality is rapidly giving way to a new, dynamic "engage and cocreate" economy. A hypercompetitive global economy is reshaping enterprises, and political and legal shifts loom.

As organizations, and indeed societies, confront this changing reality, they must ensure that they can continue to be innovative. The "who, where, what, how, and why" of innovation are in flux, across geography and economic sectors. The speed and scope of change is intensifying.

In Chapter 2, we explain how a perfect storm is gathering force and shipwrecking the old corporation in wave after wave of change. We begin by explaining how the "publish and browse," read-only Internet of yesterday is becoming a place where the knowledge, resources, and computing power of billions of people are coming together into a massive collective force. Energized through blogs, wikis, chat rooms, personal broadcasting, and other forms of peer-to-peer creation and communication, this utterly decentralized and amorphous force increasingly self-organizes to provide its own news, entertainment, and services. As these effects permeate out through the economy and intersect with deep structural changes like globalization, we will witness the rise of an entirely new kind of economy where firms coexist with millions of autonomous producers who connect and cocreate value in loosely coupled networks. We call this the collaboration economy.

Next, we take you on a tour of the collaboration economy, including seven new models of mass collaboration that are successfully challenging traditional business designs.

1. The journey begins with the "Peer Pioneers"—the people who brought you open source software and Wikipedia while demonstrating that thousands of dispersed volunteers can create fast, fluid, and innovative projects that outperform those of the largest and best-financed enterprises.

2. "Ideagoras" explains how an emerging marketplace for ideas, inventions, and uniquely qualified minds enables companies like P&G to tap global pools of highly skilled talent more than ten times the size of its own workforce.

3. The "Prosumers" takes you through the increasingly dynamic world of customer innovation, where a new generation of producer consumers considers the "right to hack" its birthright. This is good news.

4. The "New Alexandrians" will bring you up to speed with a new science of sharing that will rapidly accelerate human health, turn the tide on environmental damage, advance human culture, develop breakthrough technologies, and even discover the universe—all the while helping companies grow wealth for their shareholders.

5. "Platforms for Participation" explains how smart companies are opening up their products and technology infrastructures to create an open

stage where large communities of partners can create value, and in many cases, create new businesses.

6. The "Global Plant Floor" shows how even manufacturing-intensive industries are giving rise to planetary ecosystems for designing and building physical goods, marking a new phase in the evolution of mass collaboration.

7. The "Wiki Workplace" wraps up the journey with a look at how mass collaboration is taking root in the workplace and creating a new corporate meritocracy that is sweeping away the hierarchical silos in its path and connecting internal teams to a wealth of external networks.

For individuals and small businesses this is an exciting new era—an era where they can participate in production and add value to large-scale economic systems in ways that were previously impossible. For large companies, the seven models of mass collaboration provide myriad ways to harness external knowledge, resources, and talent for greater competitiveness and growth. For society as a whole, we can harness the explosion of knowledge, collaboration, and business innovation to lead richer, fuller lives and spur economic development for all.

Take heed. Whenever such a shift occurs, there are always realignments of competitive advantage and new measures of success and value. To succeed in this new world, it will not be enough—indeed, it will be counterproductive—simply to intensify current policies, management strategies, and curricular approaches. Remaining innovative requires us to understand both the shifts and the new strategy agenda that follows. We must collaborate or perish—across borders, cultures, disciplines, and firms, and increasingly with masses of people at one time.

2. THE PERFECT STORM

How Technology, Demographics, and Global
Economics Are Converging for the First
Category 6 Business Revolution

When thirty-five-year-old science-fiction writer Cory Doctorow gets up in the morning he likes to start his daily routine with a period of uninterrupted writing. October 31, 2005—apart from the fact that it was Halloween—was no typical day. When Doctorow flicked on his computer he was alerted to a report from a computer security researcher named Mark Russinovich. Sony-BMG had been caught covertly installing computer-crippling digital rights management (DRM) software onto the PCs of millions of music fans. Nothing gets Doctorow more incensed than corporate abuses of technology, and Sony-BMG was about to get an earful.

For Sony-BMG this was a highly unfortunate state of affairs. Doctorow is a coeditor of Boing Boing: A Directory of Wonderful Things, one of the most popular and influential organs on the Web. With an audience of 750,000 readers daily and growing, Boing Boing's readership now eclipses most mainstream media outlets (the *Wall Street Journal* is at 2.5 million and falling), and the self-described "activist, blogger, public speaker, and technology person" has no qualms about using the well-trafficked weblog as an outlet to raise hell. Sony's DRM missteps might have been overlooked as a poorly judged technical debacle. But with Doctorow fanning the flames, it burst into a worldwide public relations firestorm that has cost the company dearly.

Naturally, Sony officials claim that they were only trying to protect the company's intellectual property rights. Its DRM system was designed to prevent massive copying. When customers plopped a Sony disc into their CD drive it buried software deep within the PC operating system that restricted users to three copies.

On the surface it doesn't sound so bad. But Sony's "rootkit" software installed itself surreptitiously and relayed private information about users' actions back to the company. Even worse, the cloak-and-dagger software exposed customers' PCs to viruses. By the time the CDs could be recalled, millions of PCs had been infected. Customers who tried to remove the DRM software themselves risked nuking their Windows installation. Sony-BMG downplayed the debacle, calling it rather innocuous. "Most people don't even know what a rootkit is," said Sony-BMG's president of global digital business, "so, why should they care about it?" While Sony-BMG was managing its crisis, Doctorow was wreaking havoc, dedicating acres of space on Boing Boing to documenting every humiliating misstep, retraction, and reaction, and making Sony officials look like numskulls.

When Doctorow isn't blogging, campaigning, or speaking, he's writing science-fiction novels, and like a growing numbers of creative refuseniks, Doctorow doesn't toe the myopic industry line on intellectual property rights: Fans can download his novels free from his Web site. Readers in developing nations can even resell them at a profit. It's unconventional to say the least. But for Doctorow, it's all about experimenting with a new way of gaining exposure.

He says his problem isn't piracy, it's obscurity: the risk that one's work will get lost in the vast digital wilderness of content and voices. "The Internet has lowered the search cost for finding leisure-time activities," says Doctorow. "Anyone who is trying to offer something that people do in their discretionary time needs to compete not just with other works in the same category . . . but with all the other things you might do that are just a few clicks away."

In today's information-soaked environment, writers and content creators need to find ways to permeate people's consciousness. Giving away content and building loyal relationships are increasingly part of the arsenal creators use in the battle for people's attention.

By giving his fans free e-books, Doctorow enlists his devoted readers as unpaid evangelists for his work. "I trust my readers with the electronic copies of my book," says Doctorow, "and that kinda makes me their pal." We see this all over the Internet today. Authors engage in conversational relationships with their audiences where their fans call them by their first names.

When readers make that cognitive and emotional link it is no longer just discretionary activity. It's discretionary activity with a social dimension. All of this social networking ultimately helps Doctorow sell more books through mainstream channels like Amazon.

Doctorow's publishing philosophy is also informed by a measure of technological realism. Like many in the cyber community, he believes that "bits exist to be copied." He describes business models that depend on bits not being copied as "just dumb," and equates lawmakers who try to prop up these business models to "governments that sink fortunes into protecting people who insist on living on the slopes of active volcanoes."

It's an ethic that defines what the new Web is becoming: a massive playground of information bits that are shared and remixed openly into a fluid and participatory tapestry. Having matured beyond its years as a static presentation medium, the Web is now the foundation for new dynamic forms of community and creative expression. Throw in a healthy dose of grassroots entrepreneurship and you have a potent recipe for economic revolution—a revolution that affects not just the obvious targets such as media, entertainment, and software, but is increasingly sweeping across all industries and sectors as mass collaboration makes inroads into activities ranging from science to manufacturing.

The depth and scope of this revolution is broadening because the new Web is the natural habit for a new cohort of collaborators called the "Net Generation." For them the Web is not a library—a mere information repository or a place to do catalog shopping—it's the new glue that binds their social networks. Phenomena like MySpace, Facebook, flickr, 43 Things, Technorati, and del.icio.us aren't just Web sites, they're dynamic online communities where sprawling and vibrant webs of interaction are forming. Now this new generation of youthful users is bringing the same interactive ethos into everyday life, including work, education, and consumption.

As the new Web and the Net Gen collide with the forces of globalization we are entering what might be considered a perfect storm, where converging waves of change and innovation are toppling conventional economic wisdom. As Doctorow put it, "Humanity's natural affinity for expression, communication, and entrepreneurship is coalescing with the increased penetration of Internet connections and the growing accessibility of new user-friendly collaboration tools." Throw in a turbulent and increasingly competitive global environment and you get a powerful concoction that demands deep changes in the strategy and architecture of firms.[1]

To chart this perfect storm we explore the new Web's participatory architecture. Then we pay special attention to Net Gen—the generation that will inject the culture of openness, participation, and interactivity into workplaces, markets, and communities. We wrap up with a brief tour of the global economy and the new imperatives and opportunities for collaboration.

THE NEW WEB

From the Internet's inception its creators envisioned a universal substrate linking all mankind and its artifacts in a seamless, interconnected web of knowledge. This was the World Wide Web's great promise: an Alexandrian library of all past and present information and a platform for collaboration to unite communities of all stripes in any conceivable act of creative enterprise.

Today the Net is evolving from a network of Web sites that enable firms to present information into a computing platform in its own right. Elements of a computer—and elements of a computer program—can be spread out across the Internet and seamlessly combined as necessary. The Internet is becoming a giant computer that everyone can program, providing a global infrastructure for creativity, participation, sharing, and self-organization.

How is this different from the Internet as it first appeared? Think of the first iteration of the Web as a digital newspaper. You could open its pages and observe its information, but you couldn't modify or interact with it. And rarely could you communicate meaningfully with its authors, apart from sending an e-mail to the editor.

The new Web is fundamentally different in both its architecture and applications. Instead of a digital newspaper, think of a shared canvas where every splash of paint contributed by one user provides a richer tapestry for the next user to modify or build on. Whether people are creating, sharing, or socializing, the new Web is principally about participating rather than about passively receiving information.

Surf around today and it's clear that this new culture of participation pervades the Web. Nobody hangs around in the "publish and browse" Internet anymore. Increasingly people prefer to participate in a new generation of user-fabricated communities where users engage and cocreate with their peers.

Flickr, the ever-popular social photography site in which users post, share, and comment on the photos they take, is a quintessential example of how the new Web works. Flickr provides the basic technology platform and free hosting for photos (more sophisticated services are available by subscription). Users do everything else. For example, users add all of the content (the photos and captions). They create their own self-organizing classification systems for the site (by tagging photos with descriptive labels). They even build most of the applications that members use to access, upload, manipulate, and share their content. And increasingly, users license their photos for noncommercial use so that you can find flickr photos distributed across the Web. Flickr is basically a massive self-organizing community of photo lovers that congregates on an open platform to provide its own entertainment, tools, and services.

The bottom line is this: The immutable, standalone Web site is dead. Say hello to a Web that increasingly looks like a library full of chatty components that interact and talk to one another. Increasingly, people are engineering software, databases, and Web sites so that they not only meet private objectives, but so that they can be used in ways the originators did not know or intend. This makes it very easy to build new Web services out of these existing components by mashing them together in fresh combinations.

The result is that today's most exciting and successful Web companies and communities are stitching together their own services from shared databanks and Lego-style pieces of Web software. Rather than define the user experience and publish information for people to observe, they use Web services to create platforms for people to cocreate their own services, communities, and experiences. And when they built it, people came— usually by the tens of millions. In fact, 2006 was the year when the programmable Web eclipsed the static Web every time: flickr beat webshots; Wikipedia beat Britannica; Blogger beat CNN; Epinions beat Consumer-Reports; Upcoming beat evite; Google Maps beat MapQuest; MySpace beat friendster; and craigslist beat Monster.

What was the difference? The losers launched Web sites. The winners launched vibrant communities. The losers built walled gardens. The winners built public squares. The losers innovated internally. The winners innovated with their users. The losers jealously guarded their data and software interfaces. The winners shared them with everyone.

What does the programmable Web mean for users? It means that every time you share a photo on flickr, tag a bookmark on del.icio.us, or swap comments with your friends on MySpace, you're helping to enrich the new Web. Ditto for the math professor who edits the mathematics listing on Wikipedia; the small-business person who creates an e-commerce application using Amazon as its transaction engine; or the scientist who makes a contribution to the human genome project using a Web-enabled sequencing program. Increasingly, we are all witting and unwitting coconspirators in building one massively sophisticated computer.

As growing numbers of people and companies participate in programming this platform, the opportunities to compete and collaborate in new ways will only get richer and more powerful. To get a peek at where this is going we take a brief look at four of the new Web's cutting-edge phenomena and themes: the blogosphere, collective intelligence, the new public squares, and what we call emergent or serendipitous innovation. Each theme highlights something different about how the new Web is enabling new forms of mass collaboration and what this means for business and society.

The World's Biggest Coffeehouse

Communicating with a worldwide audience has never been easy for individuals—at least until recently. "You could build a Web site," says Greg Reinacker, founder and chief technology officer for NewsGator, a company that develops applications to retrieve, organize, and serve up newsfeeds on the Internet. "But you had to be fairly technical, and it wasn't easy to publish to it ten times a day. Blogging makes that easy," he says, "and gives people a chance to talk about whatever they want, to whomever they want in the outside world."

Today the blogging phenomenon points the way to the most profound changes the new Web will wreak on the economy. Blogs have been described as the biggest coffeehouse on earth. They capture a moment-to-moment picture of people's thoughts and feelings about things happening right now, turning the Web from a collection of static documents into a running conversation. Advertisers have already tapped into this, putting out their own blogs to engage interested customers in conversation. Firms use blogs as focus groups, regularly "listening in" on what people are saying about their company or products.

Critics of the blogosphere claim the vast new wilderness of voices adds more noise to an overly saturated media environment. With over one hundred million blogs, 1.5 million blog posts daily, and a new blog created every second, you'd think they might have a point. But ultimately their critique is trite.

Increasingly, blogs (and other forms of media) are aggregated using a technology called RSS (really simple syndication). This turns the Web into something like TiVo—a flowing stream of entertainment and news choices that individual users have asked for, perhaps stripped of commercial messages.

Meanwhile, new blog search engines like Technorati and IceRocket provide increasingly sophisticated tools to search and index the blogosphere in real time. Users get an up-to-date picture of who is saying what on the issues they are concerned about. In fact, a spike in activity in the blogosphere usually indicates that something interesting is happening. So even mainstream reporters take their cues from what bloggers are saying from moment to moment.

Though the majority of blogs are not yet of a quality to compete with commercial media, they point to the increasing ease with which end users can create their own news and entertainment and bypass established sources. Hundreds of communities of interest are forming where people engage in lively exchanges of information and views around everything from knitting to nanotechnology.

The potential for blogs to become richer and more engaging will only grow as people build audio and video into their posts. Do it yourself Web television stations like YouTube are already booming. "Now anybody with a video camera can post video on their blogs and create their own TV station," says Reinacker, whose company buids tools to take that video content and put it on the end user's television in full screen. It's no longer just the print media that is in danger, but producers of commercial television, radio, and movies as well.

Your Network Is Your Filter

Don was among the first Web analysts to call it collective intelligence: the aggregate knowledge that emerges from the decentralized choices and judgments of groups of independent participants.[2] Author James Surowiecki calls it the "The Wisdom of Crowds" and traces the application of collective intelligence across domains such as science, politics, and business.[3] For us, the ability to pool the knowledge of millions (if not billions) of users in a self-organizing fashion demonstrates how mass collaboration is turning the new Web into something not completely unlike a global brain.

Smart companies are now harnessing this potential to develop powerful new business models and systems. When you shop on Amazon, for example, you benefit not just from the distributed rating system that enables customers to review books (and for those reviews to be reviewed in turn) but from a remarkably sophisticated system that searches for similarities among the purchases of all Amazon customers in order to suggest books that you are likely to enjoy. And Google is the runaway leader in search because it harnesses the collective judgments of Web surfers. Its PageRank technology is based on the idea that the best way to find relevant information is to prioritize search results not by the characteristics of a document, but by the number of sites that are linking to it.

These examples show how people can collaboratively filter through the plethora of variety and choices on the Web (and in life generally) in the absence of expert guidance. Now, thanks to the pioneering efforts of a number of Web services, the application of collective intelligence is branching out to the way we organize and classify content on the Web, using a grassroots classification system called "tagging."

Tagging harnesses a technology called XML to allow users to affix descriptive labels or keywords to content (techies call it "metadata," or data about data). *Wired* cofounder Kevin Kelly aptly describes a tag as a public annotation—like a keyword or category name that you hang on a file, Web page, or picture. When people tag content collaboratively it creates a "folksonomy," essentially a bottom-up, organic taxonomy that organizes content on the Web.

Del.icio.us is a social bookmarking service where the simple activity of tagging and storing Web links becomes the basis for learning new things and making connections to new people. Joshua Schachter wrote del.icio.us as a way to keep track of all the things he was thinking about posting to his blog. He calls the service "a way to remember in public." "The actual database," he says, "represents crystallized attention—what people are looking at, and what they're trying to remember."

Visit del.icio.us at any point during the day and you get a snapshot of what Web surfers find interesting at that moment. People who use similar tags are likely to have overlapping interests. Those shared interests provide an incentive to find out what other like-minded people are bookmarking.

Tagging is just getting started, and has already been extended to documents, photos, videos, podcasts, e-mails, blog posts—basically any type of electronic content you can imagine. It's of course not perfect: Small discrepancies like "opensource" and "open source" and words with multiple meanings (or multiple words with the same meaning) can easily diminish a folksonomy's accuracy and usefulness. But in practice, tagging copes reasonably well with these problems and, in most cases, convergence around tags happens naturally.[4]

In doing so, folksonomies are one of many examples of how social networks gravitate naturally toward norms and conventions that enhance social productivity and connectivity.[5]

The New Public Squares

When entrepreneurs first harnessed the Web as a commercial platform the world was astounded at how so many barriers to entry had seemingly vanished. Amazon could overtake Barnes & Noble. E*Trade could compete for JP Morgan's customers. Pierre Omidyar could launch eBay—a revolutionary person-to-person marketplace now worth approximately $30 billion—from his San Jose living room.

The barriers are lower still today. Unlike the first generation of companies, today's Web companies can mostly just plug-and-play. Kim Polese,

CEO of open source software integrator SpikeSource says her costs are less than a tenth of what they were six or seven years ago. Why? The ready availability of well-honed Linux software, the Apache Web server, the MySQL database, and the PHP and Perl scripting languages—collectively known as the LAMP stack—means that a lot of the essential infrastructure is free. "The ability to start a company with a server, a blog, a wiki, and a handful of new development technologies," says Tantek Çelik, chief technologist for Technorati, "has made it much easier and a whole lot cheaper to get up and running."

Josh Petersen, cofounder of the Seattle-based Robot Co-op, cut his teeth working with Amazon's personalization group. He left in 2002 to start his own business—a social networking venture called 43 Things. "What's different," says Petersen, "is that pursuing your passion is a much more viable thing. Seven people can bootstrap a company and take a decent run at becoming a top-twenty Web property."

He likens developments in the new Web to the early history of cinema. "There was a period of time where cinema was a very technical art. You practically had to be an engineer just to run a camera." As the art form evolved, directors stepped up to become storytellers who were less and less preoccupied with cinematic engineering and more concerned with crafting rich and engaging experiences. "I think something like that is happening on the Web today," says Petersen.

It's a stark contrast from the old dot-com business models that locked in customers, preventing them from moving around. These models are less sustainable now. Customers have been empowered in new ways. More to the point, they're the people creating the content. Whether one looks at flickr (where users form communities around the streams of photos they post), or 43 Things (where people form support networks around shared life goals), or Technorati (which indexes the user-generated blogosphere), the model is the same: Customers are not just consumers of content; they are the lifeblood of the business. "As users start to realize that they are creating all this content," says Çelik, "they've also realized that they ought to have control over it."

For today's new Web companies, building trust is the alternative to controlling customers. "Something really interesting happens when you trust your customers," says Çelik. "They trust you." More and more Web companies are realizing that openness fosters trust, and that trust and community bring people back to the site. Petersen considers it a new cultural orientation. "It's about making it clear that you are open and that you are not building a walled garden of content or trying to hold people hostage in any way," he says. Call them the new public squares—vibrant meeting places where your customers come back for the rich and engaging experiences. Relationships, after all, are the one thing you can't commoditize.

Emergence and Serendipitous Innovation

Another consequence of mass collaboration among people is *emergence*—the creation of attributes, structures, and capabilities that are not inherent to any single node in the network. This is an old idea. Price is the best example of an emergent characteristic: In a competitive market no one firm sets the price. Instead, they all do it collectively.

What is significant today about the notion of emergence is that we are seeing sophisticated artifacts and outcomes emerging from relatively diffuse, loosely coupled activities of collaborating agents supported by Web-based tools. Examples of this are legion: open source software creation, the blogosphere (blogs, augmented with blogrolls and RSS feeds), Google, Amazon, collaborative filtering, scientific discovery, and wikis. These have become powerful economic forces, and the basis for highly successful business models (Google being the best example).

In other words, we are moving from the concept of emergence as a consequence of raw self-organization—the idea that independent agents acting together unwittingly create some new thing (so-called "order for free")—to a recognition that self-organization can also be encouraged and even orchestrated, as Google, Mozilla, and IBM have demonstrated.

Perhaps the most powerful characteristic of the programmable Web is that it invites collaboration by design with open standards and open application programming interfaces (API) that allow separate Web sites to intermingle.[6] Startups like flickr, 43 Things, del.icio.us, and Technorati, for example, opened up their APIs as a way to crank out new features, attract users, and scale up their businesses quickly. "It comes down to a question of limited time and, frankly, limited creativity," says Techorati's Çelik. "No matter how smart you are, and no matter how hard you work, three or four people in a start-up—or even small companies with thirty people—can only come up with so many great ideas."

It's all based on a principle the new generation of Web start-ups learned from the open source software community: There are always more smart people outside your enterprise boundaries than there are inside. By opening up their APIs companies create an environment for low-risk experimentation where anybody who wants to develop on top of their platforms can do so. Çelik says there are potentially millions of developers who might just have the right combination of skills and insight to create something really valuable. "No need to send you a formal request," says Çelik. "They can just take those APIs and innovate. Then, if someone builds a great new service or capability, we'll work out a commercial licensing agreement so that everyone makes money."

We'll come back to the new economic ecosystems that are forming around popular APIs in Chapter 7, Platforms for Participation. The most important lesson to take away for the moment is that emergent phenomena tend to win in the marketplace. Ten thousand interoperating agents can often marshal more bandwidth, more raw intelligence, and more requisite variety than the largest organization. The business challenge is to form—even to foster—symbiotic relationships with emergent structures, since the fundamental nature of self-organization is that it cannot be controlled easily. But it can be steered.

Smart firms are thinking carefully about how to navigate the field of open versus proprietary technology, and how they can leverage the self-organizing power of suppliers, employees, and customers. This is particularly important in areas like innovation and knowledge management, but it applies widely across most business functions.

The Roar of Collaborative Culture

If there is one overarching principle that defines what the new Web is, it's that we are building this thing together—one blog post, podcast, and mashup after another. The Web is no longer about idly surfing and passively reading, listening, or watching. It's about peering: sharing, socializing, collaborating, and, most of all, creating within loosely connected communities.

As Socialtext (a provider of enterprise wiki software) founder Ross Mayfield likes to say, "The new Web is about verbs, not nouns."

As people individually and collectively program the Web, they're increasingly in command. They not only have an abundance of choices, they can increasingly rely on themselves. This is the new consumer power. It's not just the ability to swap suppliers at the click of a mouse, or the prerogative to customize their purchased goods (that was last century). It's the power to become their *own* supplier—in effect to become an economy unto themselves. No matter where one looks these days there is a powerful new economy of sharing and mass collaboration emerging where people peer produce their own goods and services. If anyone embodies this new collaborative culture, it's the first generation of youngsters to be socialized in an age of digital technologies. These youngsters are on the cusp of becoming leaders, and our research shows that this generation is different.

THE DEMOGRAPHIC WATERSHED: GROWING UP COLLABORATING

All generations in developed (and increasingly, developing) countries use the Web. Seniors, for example, have time to spend and new motives for going online—communicating with their grandchildren may be the most important. However, a new generation of youngsters has grown up online, and they are bringing a new ethic of openness, participation, and interactivity to workplaces, communities, and markets. For this reason, they merit special investigation. They represent the new breed of workers, learners, consumers, and citizens. Think of them as the demographic engine of collaboration and the reason why the perfect storm is not a flash in the pan but a persistent tempest that will gather force as they mature.

Demographers call them the "baby-boom echo," but we prefer the Net Generation, as Don dubbed them in his 1997 book *Growing Up Digital*.[7] Much of the following research we present has been updated from that book in a recent study with our colleague Robert Barnard, CEO of D-Code.

Born between 1977 and 1996 inclusive, this generation is bigger than the baby boom itself, and through sheer demographic muscle they will dominate the twenty-first century. While it is smaller in some countries (particularly those in Western Europe), internationally the Net Generation is huge, numbering over two billion people. This is the first generation to grow up in the digital age, and that makes them a force for collaboration. They are growing up bathed in bits. The vast majority of North American adolescents know how to use a computer, and almost 90 percent of teenagers in America say they use the Net. The same is true in a growing number of countries around the world. Indeed, there are more youngsters in this age group who use the Net in China than there are in the United States. This is the collaboration generation for one main reason: Unlike their parents in the United States, who watched twenty-four hours of television per week, these youngsters are growing up interacting.

Rather than being passive recipients of mass consumer culture, the Net Gen spend time searching, reading, scrutinizing, authenticating, collaborating, and organizing (everything from their MP3 files to protest demonstrations). The Internet makes life an ongoing, massive collaboration, and this generation loves it. They typically can't imagine a life where citizens didn't have the tools to constantly think critically, exchange views, challenge, authenticate, verify, or debunk. While their parents were passive consumers of media, youth today are active creators of media content and hungry for interaction.

They are also a generation of scrutinizers. They are more skeptical of authority as they sift through information at the speed of light by themselves or with their network of peers. Though they have greater self-confidence than previous generations they are nevertheless worried about their futures. It's not their own abilities that they are insecure about—it's the external adult world and how it may lack opportunity.

Research shows that this generation also tends to value individual rights, including the right to privacy and the right to have and express their own views. Throughout adolescence and later in life, they tend to oppose censorship by governments and by parents. They also want to be treated fairly—there is a strong ethos, for example that "I should share in the wealth I create." They have a very strong sense of the common good and of collective social and civic responsibility.

Further, this is the first time in human history when children are authorities on something really important. An N-Gener's father may have been an authority on model trains. Today young people are authorities on the digital revolution that is changing every institution in society.

The main tenets of *Growing Up Digital* have been borne out. However, in the last decade we learned a lot more about how the Net Generation will rewrite the rules for communities, markets, and workplaces. Let's take a brief look at each in turn.

New Shared Spaces

The Net Generation's modus operandi is networking. Young people dominate many of the huge, online communities, from Facebook to MySpace, where millions of youths socialize and collaborate to do everything from evaluating companies' products and services to providing entertainment and services of their own.

MySpace is one of today's largest social networking community. Launched in 2003 as a site for twenty-somethings to discover the Los Angeles indie music scene, it soon lowered its minimum age to fourteen. Not long after, it overtook rival friendster with superior tools and features. Today millions of young people hang out on MySpace, surpassing Yahoo, MSN, Google, and eBay in Web traffic.

The extensive sociological work of Danah Boyd, a University of Berkeley–based social scientist, provides some important insights into social networks such as Tribe, LiveJournal, MySpace, and Facebook. Her "ethnographic" accounts have a loyal following on the Web, where she has established a reputation for surfacing the deep underlying truths about the social uses of technology.

In a recent talk for the American Association for the Advancement of Science, Boyd talked about how for today's teens spending time on MySpace is about reclaiming private space. "Adults control the home, the school, and most activity spaces," said Boyd. "Teens are told where to be, what to do, and how to do it. They lack control at home, and many teens don't see home as their private space."

The new private spaces are increasingly found online, where young people gather en masse, network with peers, and make shared spaces of their own. It's like a bedroom with closed doors. Except that on MySpace, they can invite one thousand friends in.

As the physical world becomes less welcoming, online space becomes more vital and appealing. Boyd argues that teens have less and less access to public spaces. "Classic 1950s hangout locations like the roller rink and burger joint are disappearing," said Boyd, "while malls and 7-Elevens are banning teens unless accompanied by their parents. Hanging out around the neighborhood or in the woods has been deemed unsafe for fear of predators, drug dealers, and abductors."

In virtual spaces teens are increasingly free to manage their interactions, form networks, and shape their own identities. Though concerns about online predators and bullying are genuine, Boyd argues that there have been more articles published on predators than actual reported incidents online.

The heart of MySpace is the personalized profile. Members fill them with interests, tastes, and values, supplemented by music, photos, and video clips that make their profiles more appealing. Even top-drawer music and movie stars have profiles, and fans can "friend" them as well.

Boyd equates profiles on MySpace with public displays of identity. Comments from friends provide a channel for feedback and affection, in which there is an element of reciprocity. "When friends comment on someone's profile or photo," says Boyd, "they expect [their comments] to be reciprocated." Though many of these relationships are shallow, Boyd argues that the process plays an important role in how teens learn the rules of social life and cope with issues such as status, respect, gossip, and trust.

It would be easy to dismiss online social networking as another fickle youth fad. Just look at friendster, the first social networking site to attract worldwide attention. Like MySpace, it erupted in 2003, growing from a few thousand users to over one hundred million. The company couldn't keep up with the massive demands on its servers. Less than six months later, most members jumped ship. They joined MySpace, which offered more elaborate profile options and the ability to post music tracks.

Will MySpace or Facebook be any different? Won't teens just swarm to the next cool new thing? As big media companies like News Corp. and Viacom invest billions to acquire such networks they will certainly hope not. But who's to say this won't happen? In any case, the fortunes of News Corp. and Viacom are in many ways irrelevant. The key point is that online social networking is uniquely attuned to the N-Gen's cultural habits and will be part of the social fabric going forward. It signals how young people today are predisposed to connect and collaborate with peers to achieve their goals. As N-Geners self-organize into an immense online force that increasingly provides its own goods, entertainment, and services, companies risk becoming irrelevant spectators. Only the smartest and most sincere companies stand a chance of becoming meaningful participants in the networks these N-Geners are forging.

Changing the World—One Peer at a Time

The media (and parents) frequently express alarm at the shallowness of MySpace. How can you have one thousand friends, they ask? Surely most of this is superfluous. But not all social networks are for swapping photos, gawking at friend lists, or sharing music files in peer-to-peer networks.

TakingITGlobal is one of the world's best examples of how N-Geners are using digital technologies to transform the world around them. With 296,000 registered members in nearly two hundred countries, a Web site in seven languages, one could mistake TakingITGlobal for the United Nations. In a sense, you wouldn't be wrong. After all, its members rub shoulders with business and government leaders at the World Economic Forum in Davos and the World Summit on Sustainable Development. Plus the site lists over two thousand youth-initiated and -managed community action projects that tackle tough issues ranging from closing the digital divide in rural India to preventing HIV in Uganda. This United Nations is run not by senior diplomats, but entirely by young people aged thirteen to twenty-four years old.

Like MySpace and Facebook, TakingITGlobal harnesses all of the latest tools, such as blogging, instant messaging, and media sharing. But it promotes a decidedly different kind of social networking. Rather than list their favorite movie stars and music tracks, members list information about the languages they speak, the countries they have visited, and the issues that they're most concerned about. Members link to other members' profiles when they share similar interests, and those links create social connections that lead to new friendships and projects. Cofounder Jennifer Corriero calls TakingITGlobal "a platform to support collaboration among young people in developing projects, in understanding and grappling with issues, and influencing the decision-making processes, especially around those issues that are directly affecting young people."

Corriero and her partner, Michael Furdyk, started working full force on TakingITGlobal in September 2000. They got support from public and private sponsors, and by early 2001 the site was up and running and quickly gathering members. Today the site and all of its activities are coordinated by a worldwide virtual network of young people. A small Toronto-based team of fifteen manages the core organization, but Furdyk and Corriero rely on volunteers for everything else. "Hundreds of people around the world have a stake in it, have some ownership in it, and collaborate every day in making it work," says Corriero.

TakingITGlobal's next task: reforming education. Ask Furdyk why educational systems around the Western world are failing and, not surprisingly, you won't get a conventional answer. It's not all down to undisciplined students, underperforming teachers, or toothless standards. According to Furdyk, the real problem is a lack of engaging content.

Academic studies over the last thirty years show that young people's interest and enthusiasm in schoolwork has declined precipitously. "Everything else has become so engaging," says Furdyk. N-Geners who go online regularly to play video games or interact on MySpace expect better experiences in the classroom. Look at today's curriculum, though, and you won't find much interactivity. "We're still learning through reading and regurgitating," says Furdyk.

TakingITGlobal's answer is a set of tools and curricular activities that will get students collaborating with other students in other countries to complete projects, and learning through active projects that make a difference in their communities. "It's real participatory, active learning," says Furdyk. "A teacher in Canada and a teacher in Nigeria sign up, create a virtual classroom, and assign students to group research projects, while the students can blog, post artwork, and collaborate on a class wiki." With the assistance of Microsoft, the project is in classrooms across North America by 2008.

If N-Geners can transform the future of the education system, what else is next? Well, how about the entire economy, for starters.

The N-Geners As Prosumers

As a generation they are awash in options for information and entertainment. Bulletin boards, price comparison sites, blogs, and instant messaging provide easy and unprecedented consumer scrutiny of offerings and sales pitches. As N-Geners navigate the hubbub, it is not surprising that the opinions of people they know (or feel they know) strongly influence their buying decisions. Michael Furdyk says, "Our generation really doesn't trust the media and advertising as much as we trust peer-to-peer opinion and social networks." Now, for the first time since the mid-twentieth-century explosion of mass marketing, reputation and word-of-mouth have become powerful forces that marketers must harness—and they can just as easily spin out of control in unpredictable ways.

N-Geners are not content to be passive consumers, and increasingly satisfy their desire for choice, convenience, customization, and control by designing, producing, and distributing products themselves. We first got a taste of this as Napster (and later Kazaa, BitTorrent, and LimeWire) revolutionized the distribution of music, television shows, software, and movies. File sharing now accounts for half of the world's Internet traffic—much to the chagrin of Hollywood—signaling that the Net Generation is renegotiating the definitions of copyright and intellectual property. Indeed, N-Geners are not only creating new art forms, they're helping to engender a new creative and philosophical openness. The ability to remix media, hack products, or otherwise tamper with consumer culture is their birthright, and they won't let outmoded intellectual property laws stand in their way.

Pew Internet and American Life Project's survey of U.S. teenagers confirms that more than half (some 57 percent) of online teens are what the project calls "content creators." That amounts to half of all teens aged twelve to seventeen, or about twelve million youth in the United States alone, and this number looks likely to grow year after year. These content creators report having engaged in myriad activities, including creating blog or personal Web pages; sharing original content such as artwork, photos, stories, or videos; or remixing content found online into a new creation.

The most popular content-creating activities are sharing self-authored content, typically through blogs. These collections of personal thoughts, opinions, and interests represent unfiltered self-expression. Remixing content they find online into their own artistic creations is also popular. In fact, one in five Internet-using teems reports having done so. Our colleague and Stanford professor Lawrence Lessig tells us that over the past few years, for example, fans of the artists represented by the label Wind-up Records have spent at least a quarter of a million hours producing and sharing more than three thousand music videos. But these are not your garden-variety music videos.

Fans (primarily kids) use their PCs to synchronize Japanese animé art with popular music tracks to create an entirely new art form called animé music videos. According to Lessig, the three thousand animé music videos relating to Wind-up Records represent just 5 percent of the total fan-generated creations circulating on one popular site. Half a million users frequent this site, and close to thirty thousand music and animé enthusiasts contribute to it.

Wind-up Records, it turns out, squandered a brilliant opportunity to engage their customers as evangelists for their artists. The company promptly asked for the videos to be removed from the site, despite the fact that some fans' videos took up to one thousand hours to create. Unfortunately this is typical. Most companies run completely amiss when attempting to deal with the N-Gen's unbridled creativity. Smart companies, however, won't send out cease-and-desist letters; they will find creative ways to engage N-Geners in the product development and distribution process.

The opportunity to bring N-Geners (and other customers) into the enterprise as cocreators of value possibly presents the most exciting, long-term engine of change and innovation that the business world has ever seen. But innovation processes will need to be fundamentally reconfigured if businesses are to seize this opportunity. We'll discuss how in Chapter 5, The Prosumers.

The Wiki Workforce

Past generations have each brought unique characteristics to the workplace, but the high-technology adoption, creativity, social connectivity, and diversity embodied in the N-Gen genuinely differentiates it from others. As workers, this generation will transform the workplace and the way business is conducted to an extent not witnessed since the "organization man" of the 1950s.

The old corporation was strongly hierarchical, with the boss being an authority on every part of the business. As eighty million young Americans enter the workforce and marketplace they will be a powerful force for all kinds of unorthodox collaborations. "Maybe we could have a different view of management," says Michael Furdyk. "If each of us has our own areas of authority we can collaborate together in multidisciplinary teams for better results."

The Net Generation's concept of work is best described by a set of nontraditional attributes or norms we identified in our research with colleague Robert Barnard. These N-Gen norms—speed, freedom, openness, innovation, mobility, authenticity, and playfulness—can form the basis of a revitalized and innovative work culture, but they also raise tough challenges for employers seeking to adapt to new expectations. We discuss these new workplace norms at length in Chapter 9, The Wiki Workplace.

The bottom line is that the N-Gen work ethic gives this generation a leg up as inherent innovators. They are on a quest for newness. They are open to new ideas. They tend to believe in diversity in all aspects of their lives. This demographic's need for freedom will take it to uncharted territory. There is strong evidence that N-Geners will demand highly collaborative and collegial work environments that balance work and life, and most of all, value fun. Their playfulness will inject entertainment value into the workplace. And while their appetite for authenticity means that they are resistant to ill-considered attempts by older generations to "speak their lingo," companies able to adapt to the new demands of N-Gen now will gain a tremendous source of competitive advantage and innovation. Those that don't will be left on the sidelines, unable to refresh their workforces as the N-Geners flow to other opportunities.

THE COLLABORATION ECONOMY

Three historic forces underpin the perfect storm of technology, demographics, and global economics. We have already described the first two: the rise of the second-generation Internet and the coming of age of a new generation of collaborators called the Net Generation. In this final section we describe the collaboration economy, which in itself is the outcome of two converging forces: a change in the deep structures of the corporation, as companies are forced to open up their walls and collaborate with external resources to create value, and the rise of a truly global economy that demands and enables new kinds of economic cooperation and opens up the world of knowledge workers to every company seeking uniquely qualified minds to solve their problem.

Coase's Law

The idea that vast self-organized networks of knowledge producers will contribute to and even challenge the traditional corporation as the primary engine of production sounds like fantasy. So deeply embedded in the fabric of society have these lumbering industrial-age creatures become that we would scarcely recognize a world without their monopoly over production.[8]

Yet a fundamental change is occurring in the way companies orchestrate the capability to innovate and create value. Smart multibillion-dollar firms recognize that innovation often begins at the fringes. Increasingly, these hierarchical enterprises are turning to collaborative self-organizing business-web (b-web) models where masses of consumers, employees, suppliers, business partners, and even competitors cocreate value in the absence of direct managerial control. Why is this happening? It's all about the declining cost of collaborating.

Oddly enough, the story begins not in the 1990s, when the Web became widely known, but in 1937. That was the year in which a young English socialist named Ronald H. Coase published a paper titled "The Nature of the Firm." Not long before, Coase had made a visit to the United States during which he toured Ford and General Motors. The industrial behemoths left him puzzled. How, he wondered, could economists say that Stalin was mistaken to run the Soviet Union like one gigantic company when Henry Ford and Alfred P. Sloan, Jr., ran their own gigantic companies in similar ways?[9] After all, the marketplace is theoretically the best mechanism for matching supply with demand, establishing prices, and extracting maximum utility from finite resources. So why don't all individuals act as individual buyers and sellers, rather than gather in companies with tens of thousands of coworkers?

Coase argued that there were good reasons for the seemingly contradictory structure of the vertically integrated corporation. One of the main reasons has to do with the cost of information. Producing a loaf of bread, assembling a car, or running a hospital emergency room involve steps where close cooperation and common purpose are essential to producing a useful product. In day-to-day practice it was not practical to break down manufacturing and other business processes into a series of separately negotiated transactions. Each transaction would incur costs that would outweigh whatever savings were achieved by competitive pressures.

First, there would be search costs, such as finding different suppliers and determining if their goods were appropriate. Second, there would be contracting costs, such as negotiating the price and contract conditions. Third, there would be coordination costs of meshing the different products and processes. Coase called these "transaction costs." And the upshot was that most corporations concluded it made the most sense to perform as many functions as possible in-house.

All this leads to what we and our colleagues call "Coase's law": A firm will tend to expand until the costs of organizing an extra transaction within the firm become equal to the costs of carrying out the same transaction on the open market. As long as it is cheaper to perform a transaction inside your firm, keep it there. But if it is cheaper to go to the marketplace, do not try to do it internally.[10]

How has the Internet affected Coase's law? Strictly speaking, the law remains as valid as ever. But the Internet has caused transaction costs to plunge so steeply that it has become much more useful to read Coase's law, in effect, backward: Nowadays firms should shrink until the cost of performing a transaction internally no longer exceeds the cost of performing it externally. Transaction costs still exist, but now they're often more onerous in corporations than in the marketplace.

Take another look at Coase's three kinds of collaboration costs. Henry Ford and Alfred P. Sloan had to physically seek out promising suppliers, inspect their factories, and haggle over prices. In most cases, the costs of sourcing external suppliers were so high that ownership of production processes across the automotive value chain was warranted, even if individual links were thereby shielded from market disciplines. Ford Motor's infamous River Rouge plant, which drew raw rubber and steel into one end and pushed finished cars out the other, was the quintessential example of this. Today's automotive firms type "axle" or "window glass" into any number of industry exchanges, and negotiate the price on the Web. If they want, they can check a supplier's trustworthiness—just scan through the treasure trove of analytic services available online.

How about contracting and coordination costs? Need steel from China, rubber from Malaysia, or glass from Wichita, Kansas? No problem. Online clearinghouses for each of these products enable purchasers to contract for price, quality, and delivery dates with a few clicks of a mouse. An executive can even track each shipment on a virtual map that shows its precise location at any point in the journey.

Coase's law, which once provided such a neat explanation for the development of the gigantic corporation, now explains why traditional corporations are finding themselves thrust aside by an entirely new kind of business entity.

The Business Web

That new entity is the business web, or the b-web. B-webs are clusters of businesses that come together over the Internet. While each company retains its identity, the companies function together, creating more wealth than they could ever hope to create individually.

Business webs are predicated on a new kind of interenterprise collaboration. They are not merely a refined alternative to the old corporation, like a new brand of cigar at a yacht club for contented executives. B-webs amount to a tidal wave crashing down on the club itself.

Until just a couple of years ago, a small number of industrial-age corporations, such as Sony and PolyGram, controlled the music business. They paid scant attention to MP3, a freely available standard for the compression and transmission of digital audio. Yet under their noses, an MP3 b-web was taking shape. The nascent b-web of the late 1990s combined content companies such as MP3.com; manufacturers of the Rio MP3 player (a predecessor of the iPod); distribution technologies like Napster, developed by nineteen-year-old Shawn Fanning; and the countless teenagers who were high on music, low on cash, and convinced that "information wants to be free."

Before the Internet it would have been virtually impossible for so many disparate groups of people to find each other—the search costs would have been too high. On the Internet, search costs hardly exist. One teenager discovers MP3.com while surfing the Web and sends an e-mail to his friends to tell them about the site, and a b-web begins to form.

By 2000, when the music industry finally noticed it, the MP3 b-web had reached critical mass—tens of thousands of music files had become available for downloading over the Net—and Napster alone, record companies said, had cost them $300 million in lost sales.

Today the music industry finds itself launching one expensive lawsuit after another against Internet music companies, and even against its own customers. Many labels are desperately struggling to prop up the old way of doing business while straining to figure out the new.

The rise of business webs does not mean that companies will never vertically integrate by moving knowledge and capabilities inside their boundaries. Every company needs to constantly adjust their boundaries to meet ever-changing demands and opportunities. This means that boundary decisions are at the heart of business strategy. What capabilities should be in and what should be out? If we collaborate in a business web how do we divide the spoils? How do we make sure we harvest our share of the value?

Apple has done a good job of building a business web for music—one in which they are capturing much of the value. Apple took chips (made by Motorola and others); placed them in an iPod (manufactured by Foxconn); hired developers to program iTunes; and provided access to digital music (created by publishers and artists). The result was a unique and enormously successful music experience for customers. But most of the money in this b-web goes to record labels, as your $300 iPod may have $3,000 worth of songs on it. To capture more of the value it makes sense for Apple to become a music publisher—so get ready for Apple Records redux. In fact, Apple is offering a variety of music that fans can only get through iTunes.

With so much at stake it's easy to see why business-web collaboration is not exactly a love-in. Joining or leading a business web often involves jockeying between players for the right division of labor and value. And it's not unusual for the company who orchestrates the b-web to be worried about problems such as opportunism among its partners. Some of the computer and electronics companies we just mentioned worry about the Taiwanese electronics manufacturing company Foxconn—a company that has grown to become the biggest in the industry. Foxconn has great prices, achieved through its massive manufacturing facilities in China. But as one computer executive told us, "I worry that I'll wake up one morning and they'll have moved up the food chain to become my competitor rather than supplier. Their low prices are like a drug—I love them. But I don't trust them either."

As what could be called "the enterprise 2.0" sets sail in the perfect storm it will need to navigate these tough issues. Meanwhile, the deepening process of globalization will heighten the need for b-web leaders to act globally.

The Global Playing Field

Of course, when it comes to constructing global business webs, the topic on most business managers' minds is the rise of India and China. Much of the mainstream business press is focused on outsourcing and offshoring. While this is understandable, it is ultimately too narrow. The issue is much bigger than both. We are witnessing the reweaving of the social, political, and economic fabric that binds our planet, with long-term consequences that are as or more profound than those of the industrial revolution.

Mapping the Global Future, a report from the National Intelligence Council, concludes: "In the same way that commentators refer to the 1900s as the 'American Century,' the 21st century may be seen as the time when Asia, led by China and India, comes into its own." China is already a manufacturing powerhouse; India is now the global office. Together, the "Chindia" region's potential of huge domestic markets—encompassing a third of humanity—cheap, highly-skilled labor, and governments pursuing capital-friendly policies have led many to conclude that the world is at a tipping point in history.[11]

Some have appropriately raised doubts as to whether these countries have the resources necessary to lead the world in advanced scientific research or deliver genuinely innovative products and services. Both countries, for example, will have to dedicate a sizable proportion of their human and financial capital to supplying infrastructure, health care, social security systems, and better schooling if they expect to remain competitive in a knowledge economy.[12] Meanwhile, Madhav Bhatkuly at TCI New Horizon Fund in India says, "China and India have too much low-hanging fruit that they can pick purely on the basis of low cost for them to really focus on technology that sends man to the moon!"[13]

These points have merit, but it is a grave mistake to assume that capabilities and advantages are permanent. Leading Asian juggernauts are arguably leaping over the structural and organizational hurdles they face, and skeptics often underestimate their capacity and eagerness for rapid learning.[14]

ICICI Bank, a Mumbai-based outfit, has become India's second largest retail bank, from a standing start, in under ten years. The bank leads in every retail product market that it targets. Now ICICI is knocking on the doors of global banks' home markets, with brand-new subsidiaries in Canada and the U.K., each growing by over $1 million in assets every day.

ICICI's success is due partly to the self-service proposition (customers perform their own transactions in self-serve e-lobbies), and partly to low Indian labor costs. But ICICI's costs are also low because its innovative technology systems—generally servers running open source software—cost less than one-tenth of the business infrastructure employed by the average Western financial services company.

In China, "innovation cities" are emerging across the country, where thousands of intermingling companies leverage technology, low-cost structures, and physical proximity to destroy their worldwide competition. These amorphous business webs attack industries from motorcycles to mobile phones with a deft ability to infuse their own innovations into knockoffs of Western products. Not only do they produce them at a fraction of the cost, they often do so without a recognizable company appearing on the radar screen of established competitors.

The truth is that China and India have come a long way very fast, and both are determined to go much further. Globalization now extends from labor-intensive semiskilled manufacturing all the way to high-tech design and services. This trend is unlikely to diminish. Even health care is now an Indian export. Leading centers in India provide open heart surgery for less than a quarter of the price in Europe or the United States.

It is unrealistic—indeed reckless—for the United States or Europe to think that they will easily dominate the high-tech economy as they have in the past. U.S. and European firms will soon compete in a world in which they produce maybe only one of every four or five major inventions; in which their wages and health care costs are several times higher than those of emerging competitors; and in which both the largest technological work-forces and markets are in other nations.

The future, therefore, lies in collaboration across borders, cultures, companies, and disciplines. Countries that focus narrowly on "national goals" or turn inward will not succeed in the new era. Likewise, firms that fail to diversify their activities geographically and develop robust global innovation webs will find themselves unable to compete in a global world. Effectively, it's globalize or die.

Just ask Canadian electronics manufacturer Celestica. Like many Western-based firms in the electronics manufacturing services industry, the company has undergone a massive restructuring, whereby companies are shifting operations to Eastern Europe, Latin America, and (primarily) Asia. Celestica has shifted its plants from 80 to 20 high- to low-cost geographies to 20 to 80. "If we hadn't made this wrenching change in the last few years," says CEO Steve Delaney, "I honestly don't think we'd be around today."

Globalization doesn't just lead to rationalized supply chains. It also affects the geography of innovation. The world provides a diverse set of environments for innovation, depending on factors like technology infra-structure, country-specific skills, income levels, and competitive dynamics. Innovative multinational firms have long tapped these diverse environments for new ideas. Two things have changed. The first is that the spectrum of technology maturity is very broad, with the United States often somewhere in the middle rather than at the leading edge. South Korea is far ahead of the United States in broadband Internet availability. The same can be said of high-speed mobile phone usage, making South Korea an exceptionally in-teresting place to look for consumer-oriented IT innovation. The second change is the ease with which R&D teams can now collaborate across geog-raphies using Web-based tools and accelerated product development cycles.

Western pharmaceutical giants such as Eli Lilly and GlaxoSmithKline are tightening their belts by aggressively outsourcing R&D functions at nearly all stages of the R&D value chain: from early-stage research, where smaller firms are proving more efficacious, through to the management of clinical trials, which are notoriously expensive. An estimated 20 to 30 percent of worldwide clinical research already takes place in Asia.[15]

Many worry that the globalize or die imperative will lead to the eventual decay of the West. But recent research shows that globalization does not give us as much to fear as some people think. For instance, executives are discovering outsourcing is really about corporate growth, making better use of skilled staff, and even job creation, not just cheap wages abroad. As IBM CEO Sam Palmisano explains, "Tapping global talent pools is not about arbitrage, but about capability." Yes, labor savings from global sourcing can still be substantial. But it's small potatoes compared to the enormous gains in efficiency, productivity, quality, and revenues that can be achieved by fully leveraging offshore talent.[16] As long as we take care to assist those who lose out in the transition, globalization promises more wealth and jobs for everyone.

Even small and medium-size businesses are seeing the upside in the rise of a more open global economy. Unlike large firms, they are not deeply encumbered by large workforces, substantial investments in physical plant, or organizational rigidities. Now, with a truly global business infrastructure at hand, they can literally buy, off the shelf, practically any function they need to run a company. Everything from market research to back-office support to contract manufacturing is available on tap. And most, if not all of it, can be managed over the Web. Suddenly the small can become very large very quickly with superefficient business models that enable them to exploit world-class capabilities.

In the collaboration economy, the real advantage of global sourcing is not cost savings, but the endless possibilities for growth, innovation, and diversity. The brainpower behind the next generation of products and services will be more numerous, more diverse, and more distributed than at any time in history. Tomorrow's global marketplace will provide an abundance of opportunity and a plethora of choices and variety. And the greatest growth engines of the twenty-first century will be business webs that fuse the resources and competencies of the developed and developing worlds into unbeatable combinations. As it turns out, in the global economy no man, company, or nation is an island. And the bona fide new business rule for competitiveness, "collaborate or perish," is a global one.

RIDING THE PERFECT STORM

In this chapter we have tried to substantiate the new promise and imperatives of collaboration. We have argued that the coming together of a global platform for collaboration, a generation that grew up collaborating, and a global economy that enables new forms of economic cooperation is creating the conditions for a perfect storm that will drive deep changes in the strategy and architecture of firms. The old, ironclad vessels of the industrial era will sink under the crashing waves, while firms that create highly nimble and networked structures and connect to external ideas and energies will gain the buoyancy they require to survive. Companies that anticipate and lead these changes will win important advantages in their industries.

As we shared our perfect storm metaphor with a colleague, we were reminded that the drive toward new collaborative forms of organization is as long as human history. So while we may have started this chapter with the new Web and the Net Gen, we could have easily gone back two hundred thousand years when, according to evolutionary biologists, our species began to evolve the capacity for abstract thought and communication. And since we both love history we thought we might sum all this up by way of one last dive into the past.

Our ancestors left the first tangible evidence of our predisposition to social cooperation sixty to seventy thousand years ago in the cave paintings and primitive tools left behind by hunter-gatherer communities. We learned quickly that groups with cooperative habits were more materially successful than those in which more narrowly self-interested behavior was the norm. So around 8000 B.C., our species began to settle down. Instead of foraging for food, we began to tend crops and keep herds, storing them in small village settlements that limited our mobility. We devised a cooperative division of labor between communities and constructed social rules and institutions. Cities, armies, empires, enterprises, nation states, and social movements emerged within a mere wink of evolutionary time.

Today humans engage in an elaborate pattern of task sharing that is well beyond individual comprehension. Most of us obtain a substantial share of our sustenance from strangers—people to whom we are not related by blood or marriage. Our systems of production, trade, and finance are truly global, operating with real-time information and harnessing re-sources from around the globe. We can exploit the capabilities of large numbers of specialized producers who come together to cooperate on projects while separated by time and space. And we have accumulated a staggering body of knowledge given the relatively short time span in which we have devoted significant resources to thinking.

Now, having largely mastered the productive challenges of our physical environment, we find ourselves confronting the opportunities of the cerebral environment, an increasingly virtual world of knowledge, media, and entertainment; a world girdled by information involving billions of connected individuals; a world where anyone can plug-and-play and where collaboration between diverse entities is the modus operandi of the day. We call it the world of "wikinomics"—in which the perfect storm of technology, demographics, and global economics is an unrelenting force for change and innovation.

What are business managers to make of this perfect storm? How can companies adapt? The next seven chapters will outline new strategies and business models that companies are using to innovate and rewrite the rules of competition. Peer production, ideagoras, prosumer communities, the new Alexandrians, open platforms, global plant floors, and wiki workplaces each represent new ways for companies to harness mass collaboration for better innovation and growth. We hope you enjoy the ride.

3. THE PEER PIONEERS
Software Was Just the Beginning

On July 7, 2005, at 8:50 A.M., the city of London, England, came to a standstill as four synchronized bombs exploded in its transportation system. Eighteen minutes later, as media outlets scrambled to cover the story, the first entry appeared in Wikipedia, a free online encyclopedia that anyone can edit. Morwen, a wiki enthusiast from Leicester, England, wrote: "On July 7, 2005, explosions or other incidents were reported at various London Underground stations in central London, specifically Aldgate, Edgware Road, Kings Cross St Pancras, Old Street, and Russell Square tube station. They have been attributed to power surges."

Within minutes, other community members were adding additional information and correcting her spelling. By the time North Americans woke up, hundreds of users had joined the fray. By the end of the day, over twenty-five hundred users had created a comprehensive fourteen-page account of the event that was much more detailed than the information provided by any single news outlet.[1] In doing so, they provided a potent example of the power of Wikipedia, while demonstrating that thousands of dispersed volunteers can create fast, fluid, and innovative projects that outperform those of the largest and best-financed enterprises.

Wikipedia is an example of peer production, a new way of producing goods and services that harnesses the power of mass collaboration. Though peer production is typically associated with the dispersed teams of open source programmers who work on free software projects such as Linux, the real story today is that Linux is an economic powerhouse. Linux-related hardware and services produce billions of dollars of revenue annually, and now IBM, Motorola, Nokia, Philips, Sony, and dozens of other companies are dedicating serious resources to its development.

At the time of writing over one hundred million users of set-top cable boxes, TiVos, Motorola Razrs, and other home appliances use Linux, and over a billion people use it indirectly whenever they access Google, Yahoo, or myriad other Web sites. If you drive a BMW, chances are it's running Linux. Hardware vendors sell over $1 billion worth of Linux servers per quarter—and IBM's hardware business alone was over $5 billion in 2008. Linux is growing eight times faster than the server market overall, and it has been adopted by big users like the People's Republic of China—a fairly large organization.

Linux and Wikipedia raise a number of important questions that are central to this chapter. If thousands of people can collaborate to create an operating system or an encyclopedia, what's next? Which industries might be vulnerable to peer production? And which firms might be positioned to benefit? Should smart managers try to bury the phenomenon, the way the music industry has tried to quash file sharing? Or can they learn to exploit the creative talents that lie outside their boundaries, the way IBM and others have harnessed open source software? And if, as an entrepreneur, I open source my product or technology, how do I make money from an asset that I no longer own or control directly?

Questions that might have sounded academic a few years ago are suddenly vitally important. Peer production is emerging as an alternative model of production that can harness human skill, ingenuity, and intelligence more efficiently and effectively than traditional firms. The way companies address peer production will shape the future of industry and affect their very chances for survival.

Companies will need a strategy to address peer production that takes into account both the threats and opportunities for their businesses. So in this chapter we dig deep to retrace the journeys and uncover the lessons of the "peer pioneers"—the people who brought you open source software and Wikipedia. We explain how peer production works and what these early examples tell us about the likely trajectory of the phenomenon. We dispel the myth that peer production will only take wealth away from the economy and thereby erode the ability to make profits. Then we show how companies can harness peer production for profit and lay out the key business benefits.

If there is one thing to take away from this chapter it's that treating peer production as a curiosity or transient fad is a mistake. It's not just online networking. Or, as Google CEO Eric Schmidt might say, peer production is about more than sitting down and having a nice conversation with nice objectives and a nice attitude. It's about harnessing a new mode of production to take innovation and wealth creation to new levels.

The time to address peer production is now. Barriers to entry are vanishing and the trade-offs that individuals make when deciding to contribute to projects and organizations are changing, creating opportunities to dramatically reconfigure the way we produce and exchange information, knowledge, and culture. Companies that recognize, address, and learn to tap peer production will benefit, while those that ignore and resist will miss important opportunities for innovation and cost reduction, and may even go out of business.

A NEW MODE OF PRODUCTION

Before we launch into the stories of the peer pioneers, a few words about peer production are in order. First of all, what is it and how does it work?

In its purest form, it is a way of producing goods and services that relies entirely on self-organizing, egalitarian communities of individuals who come together voluntarily to produce a shared outcome. In reality, peer production mixes elements of hierarchy and self-organization and relies on meritocratic principles of organization—i.e., the most skilled and experienced members of the community provide leadership and help integrate contributions from the community.

In many peer production communities, productive activities are voluntary and nonmonetary. They are voluntary in that people contribute to these communities because they want to and because they can. No one orders a worker to post an article to Wikipedia or to contribute code to the Linux operating system. They are nonmonetary because most participants don't get paid for their contributions (at least not directly), and individuals determine if, what, and how much they want to produce. Just because people don't get paid to participate in peering does not mean, however, that they do not benefit from their participation in other ways. We'll come back to this point in a moment.

People have been collectively raising barns since time immemorial, you say, so why is this new and different?

For starters, the economics of production have changed significantly as we have moved from an industrial to an information-based economy.[2] In the industrial economy, for example, most opportunities to make things that were valuable and important to people were constrained by the high costs of making them. If you wanted to publish a mass-circulation newspaper you needed a printing press and a physical distribution infrastructure for delivering your papers door-to-door. Simply wanting to do this was not a sufficient condition to make it happen. You needed financing to obtain the physical capital. And, in order to get an acceptable return on invested capital, you needed to orient production toward the market (i.e., you needed to sell subscriptions).

Today, billions of connected people around the planet can cooperate to make just about anything that requires human creativity, a computer, and an Internet connection. Unlike before, where the costs of production were high, people can collaborate and share their creations at very little cost. This means that individuals needn't rely on markets or capital-intensive firms to make or trade all of the goods and services they desire. In fact, a growing proportion of the things we value (including newspapers) can now be produced by us or in cooperation with the people we interact with socially—simply because we want to.

This sounds like a potential threat to business. But, in fact, it's an opportunity for companies to learn how to harness this creative potential in their businesses. Sun Microsystems, for example, was once seen as hostile to Linux and other open source software, for reasons that will become clear later in the chapter. Now Sun is publishing the inner workings of its high-end SPARC microprocessors under the same permissive licenses that govern Linux. "We'll give the secret sauce to a generation of students and academics as well as potential competitors and manufacturers in the U.S., China, India, and Eastern Europe," says Jonathan Schwartz, the company's chief executive officer. It's no gag, however. Sun sees this as a way to broaden its community of collaborators, boost support for its product, and create spin-off opportunities for Sun and its partners.

So, how can loose networks of peers possibly assemble goods and services that compete head-to-head with those of a large, deep-pocketed company?

For one, peering taps into voluntary motivations in a way that helps assign the right person to the right task more effectively than traditional firms. The reason is self-selection. When people voluntarily self-select for creative, knowledge-intensive tasks they are more likely than managers to choose tasks for which they are uniquely qualified. Who, after all, is more likely to know the full range of tasks you are best qualified to perform— you or your manager?

As Linus Torvalds, the originator of Linux, says: "People just self-select to do projects where they have expertise and interest." As long as communities have mechanisms for weeding out weak contributions, then large, self-selecting communities of people in constant communication have a higher probability of matching the best people to the right tasks than a single firm with a much smaller set of resources to work with. This applies to domains such as research and engineering as much as it does to software, education, and entertainment.

The California Department of Education, for example, thinks it can harness the insights and spare time of its teachers to make high-quality educational materials available to every aspiring student while saving local taxpayers over $400 million every year. The California Open Source Textbook Project runs on the same software that powers Wikipedia. It's already running a pilot program to create a world history text for tenth-grade history classes. By doing so it is joining tech companies like IBM and Sun and top universities such as MIT in developing free open source educational materials that anyone can use and large communities of educators can improve.

This highlights another important feature of peer production—the traditional notion of property rights is inverted.[3] Traditional forms of intellectual property confer the right to exclude others from using or distributing a creative work. Peer production is more or less the opposite. Communities of producers typically use "general public licenses" to guarantee users the right to share and modify creative works provided that any modifications are shared with the community. By opening up the right to modify and distribute, these open source licenses allow larger numbers of contributors to interact freely with larger amounts of information in search of new projects and opportunities for collaboration.[4]

Thus, by doing away with the overhead of contracts and negotiations, and enabling participants to work on any project they find interesting, peer production can be more efficient at allocating resources. But what motivates people to freely contribute their time and talents to projects like Linux and Wikipedia?

People participate in peer production communities for a wide range of intrinsic and self-interested reasons. For example, when we asked Linus Torvalds why programmers devote huge parts of their lives to building Linux without any direct monetary compensation, he replied: "If you were a software engineer you wouldn't even ask that question. For an engineer, when you solve some technical problem, the hairs just stand up on the back of your neck, it's so exhilarating. That feeling is what drives me." Basically, people who participate in peer production communities love it. They feel passionate about their particular area of expertise and revel in creating something new or better.

But the motivations for participating are ultimately much more complex than fun and altruism. People who work on Linux during their spare time are usually employed in some other facet of the industry. Participating in Linux gets them experience, exposure, and connections, and if they're good, they can earn status within the community that could prove to be highly valuable in their careers. What's more, a growing number of people are paid to participate in Linux by the companies they work for. In fact, IBM and Intel are two of the largest contributors to Linux in terms of manpower. And yes, even Linus Torvalds makes his living coordinating Linux development through the nonprofit consortium Open Source Development Lab. So, how far can peer production go?

Peering works best when at least three conditions are present: 1) The object of production is information or culture, which keeps the cost of participation low for contributors; 2) Tasks can be chunked out into bite-size pieces that individuals can contribute to in small increments and independently of other producers (i.e., entries in an encyclopedia or components of a software program). This makes their overall investment of time and energy minimal in relation to the benefits they receive in return. And, finally, 3) The costs of integrating those pieces into a finished end product, including the leadership and quality-control mechanisms, must be low.

Even when these conditions are present, peer production still faces obstacles. Communities need systems of peer review and leaders who can help guide and manage interactions and help integrate the disparate contributions from users. They also need to design rules for cooperation, cope with free riders, and figure out ways of motivating and coordinating collective action over long periods of time. Despite such obstacles, open, self-organized communities of producers seem to work—sometimes to miraculous effect.

But in the cold light of day, it's no miracle at all. Peer production works because: the new economics unleashed by technology have permanently altered the costs and benefits of producing information and collaborating; it is more efficient than firms or markets at allocating time and attention for certain tasks; it is really good at attracting a more diverse, broadly dispersed talent pool than individual firms can muster; and contributors enjoy the freedom and experience of peer production. In short, peer production works because it can.

THE ENCYCLOPEDIA THAT ANYONE CAN EDIT

Wikipedia founder Jimmy Wales is onto something big. Or should we say huge? Wikipedia, after all, is now the largest encyclopedia in the world, offered for free, and created entirely by volunteers on an open platform that allows anyone to be an editor. It's amazing that Wikipedia exists at all, let alone that it includes over four million articles in over two hundred languages. It has become one of the most visited sites on the Web. It represents the future of publishing, and every company that produces information—from publishers to data providers—should be scared.

It's not just its size or popularity, but also the way Wikipedia has evolved that makes it unique. Thousands of Web users volunteer their time and knowledge to help fulfill the community's goal of providing every person in the world with a high-quality encyclopedia in their native language. "Imagine a world in which every single person on the planet is given free access to the sum of all human knowledge. That's what we're doing," says Wales.

Built on Web software called "wiki" (Hawaiian for quick), Wikipedia allows multiple users to create and edit the same Web page. It is built on the premise that collaboration among users will improve content over time, in the way that the open source community steadily improved Linus Torvalds's first version of Linux.

Wales first ventured into the world of encyclopedic content in 1998, when he established Nupedia with former employee Larry Sanger. Like Wikipedia, Nupedia allowed anyone to submit articles and content. Unlike Wikipedia, it was a centralized, top-down hierarchy: paid academics and topic experts followed a laborious seven-step process to review and approve content. One year and $120,000 into the project, Nupedia had only published twenty-four articles, and Wales decided to scrap it.

One of Wales's employees then introduced him to the wiki, a concept invented by Ward Cunningham in March 1995, and Wales started again with a much more open way of organizing the site that would allow anyone with the inclination to participate. In the first month, Wikipedia published two hundred articles, and in the first year the total reached eighteen thousand.

Today, Wikipedia is written, edited, and almost continuously monitored by an ever-growing number of online volunteers. Of the one million registered users, roughly one hundred thousand have contributed ten or more entries. Then there's the hard-core group of about five thousand Wikipedians who gladly accept responsibility for the large variety of tasks that keep Wikipedia humming.

To some it remains a mystery why people volunteer to peer produce Wikipedia. Wales just shrugs. "Why do people play softball? It's fun, it's a social activity." Wikipedia also attracts a lot of subject matter experts. They're passionate about their topics, and they want the world to know about it. Then there's the charitable mission. "We are gathering together to build this resource that will be made available to all the people of the world for free," says Wales. "That's a goal that people can get behind."

Wikipedia's Division of Labor

Hearing Jimmy Wales talk about his fellow Wikipedians gives one a deeper appreciation for the highly specialized division of labor that has evolved to nurture the site's growth. While most people go to Wikipedia to read the content, there's a ton of administrative work that goes on behind the scenes. This includes duties like: administering pages, developing software, finding copyright-free photos, moderating conflicts, and patrolling for vandalism. With only five paid staffers, volunteers perform most of it.

Wales suggests we examine the history of one entry on Wikiepdia as an example, so Don proposed the Bernese mountain dog (his family had just gotten one—go to flickr if you want to see Arnold as a puppy). "I see Trysha reduced the image size, so she's coming around and basically adjusting the format and making sure that the article is consistent with others," says Wales. "Here's Elf. She's adding interlanguage links from this article to the Dutch version of Wikipedia. That's one of the things that she does a lot."

Elf, a self-confessed Wiki addict, is also a dog fanatic. When she's not busy with her full-time occupation as a competitive dog trainer, she's on Wikipedia, where she admittedly spends "too much time" carefully curating the thousands of the articles that Wikipedians have contributed on various dog breeds. Like thousands of other Wikipedians, Elf voluntarily maintains a watch list of hundreds of articles and photos that she monitors every time a change is made. She helps ensure the accuracy of editorial changes and can quickly remedy any vandalism.

Unlike a traditional hierarchical company where people work for managers and money, self-motivated volunteers like Elf are the reason why order prevails over chaos in what might otherwise be an impossibly messy editorial process. Wales calls it a Darwinian evolutionary process, where content improves as it goes through iterations of changes and edits. Each Wikipedia article has been edited an average of twenty times, and for newer entries that number is higher. Despite the huge number of users, Wales estimates that over 50 percent of edits are made by less than 1 percent of users, a clear sign that amid the chaos lies a small but committed group of regular users.[5] On occasion, "edit wars" break out, in which users repeatedly reverse each other's changes. In these rare cases, a Wikipedia staffer makes the final judgment.

Growing Pains

It's not surprising that Wikipedia is not perfect. For all its attempts to manage quality, the collaborative production model engenders some risks. Case in point: In May 2005, an anonymous Wikipedia user created an almost entirely fictional article on former *USA Today* editorial director John Seigenthaler, Sr. "John Seigenthaler Sr.," it read, "was the assistant to

Attorney General Robert Kennedy in the early 1960s. For a brief time, he was thought to have been directly involved in the Kennedy assassinations of both John, and his brother Bobby. Nothing was ever proven."

For the next four months, any Wikipedia user (or any user of the fifty-two sites that mirror Wikipedia's content) looking up Seigenthaler would have read the erroneous biography. Seigenthaler later called it Internet character assassination.

The incident exposed the most obvious weakness of the Wikipedia model: Anybody can claim to be an expert on any subject. And while the site is designed to empower users to self-police, the publicity around this incident hurt the site's credibility. Wales has since introduced a policy that prevents unregistered users from creating new articles on Wikipedia, but savvy Net users can and will circumvent the policy by registering with bogus names and a free e-mail account. Seigenthaler later asked Wales if he had any way to discover who wrote the entry. Wales's answer: "No, we don't." But Wikipedia has recently moved to freeze entries like George W. Bush that are obvious magnets for vandalism and tampering.

Seigenthaler is hardly the only detractor. Academic critics argue that episodic vandalism and uneven quality undermines Wikipedia's authority as a scholarly resource. Expertise is by no means shunned on Wikipedia, but "credentialism" is clearly discouraged. A Ph.D. in astrophysics may just as well find him- or herself arguing over the nature of the universe with an eager high school student (or worse, an astrologer) as with a peer with equivalent training.

This looseness leads some professors to discourage their students from using the free encyclopedia as a reference tool. Some worry that students could make up an entry and then cite it as the source! Other academics are more optimistic, and some are regular contributors. Take Matt Barton, a professor of English at St. Cloud State University in Minnesota, who's recently taken to Wikipedia to build a living, breathing resource on English rhetoric, its history, uses, and meaning.

"I could sit down and take days, weeks, even months to find all the terms," says Barton, "but with Wikipedia, I can start the list with three or four definitions and then kick back and let the community chip in a little." Barton figures that the more eyes on his work the better, so he blogs about his work and invites students and peers to get involved. "I might make a mistake that I'm not seeing, so having those people watching over it is a good thing."

There is a growing contingent of scholars like Barton that see the value of a dynamic, evolving body of knowledge, even if they also encourage their students to consult additional texts. Wales himself encourages students to consult other sources when doing scholarly research. But he works tirelessly to instill the values of "neutral, high-quality information" in the modus operandi of the site, and believes that fostering a larger creative community of contributors and editors is the route to higher quality. Controversial entries that fail to meet Wikipedia's standards can be edited, locked, or nominated for deletion by users.

As for inaccuracies, so-called expert sources may not be as justified in making claims of authority as they think. Indeed, unfavorable comparisons between Wikipedia and *Encyclopedia Britannica* may not be based in much fact. *Nature* magazine's comparative analysis of forty-two science entries in both showed a surprisingly small difference: Wikipedia contained four inaccuracies per entry to *Britannica*'s three.[6]

Britannica has disputed this finding, saying that the errors in Wikipedia were more serious than the *Britannica* errors, and that the source documents for the study included the junior version of the encyclopedia as well as the *Britannica* yearbooks.[7]

Unfortunately for *Britannica*, its complaints really miss the point—errors cited on Wikipedia have long since been fixed, while the *Britannica* errors remain. In the same way that open source programmers swarm together to identify and fix bugs, Wikipedians can easily catch errors and set the record straight. According to an MIT study, an obscenity randomly inserted on Wikipedia is removed in an average of 1.7 minutes.

Wikipedia will no doubt always have its critics. Robert McHenry, former editor-in-chief of *Britannica*, complains that despite the process of iterative edits, the quality of Wikipedia entries is still "what might be expected of a high school student."[8]

True, Wikipedia's openness leaves it vulnerable to inaccuracies, edit wars, and vandalism. But its openness is also the reason why it's constantly growing, adding new entries, covering new niches, and always reviewing and updating facts. It taps an almost infinite wealth of talent, energy, and insight that far exceeds what *Britannica*'s closed model can muster. Over time

Wikipedia will likely gravitate toward a model where the community—
and perhaps even an editorial board that is representative of Wikipedia's
constituents—can help accredit articles and verify sources more effectively,
creating greater reliability and trust in the content.

Treading Lightly

For now, Wales prefers to tread lightly when it comes to implementing
top-down controls, and fears that exerting too much control too quickly
will kill the community spirit. "We could be draconian about how we police the site," says Wales, "but that's like throwing everyone in jail for any
minor infraction. We want to get out there and clean up the park so people
don't feel that they live in a slum and can break windows if they want to.
We'd rather try to build a healthy, positive environment so people feel
positively inclined to contribute in a constructive way."

So far, it's working. This largely bottom-up process of improving the
site and its processes keeps Wikipedia growing at an astonishing rate. An
average of nearly two thousand new English-language articles are posted
every day on every imaginable subject. That adds up to 730,000 new articles each year. Today the English-language version of Wikipedia alone
boasts over 3 million articles. At some point, Wales will lock in a Wikipedia
1.0, at which point a body of high-quality entries will be frozen. For as long
as the world is changing, however, there will be plenty of room for new
content.[9]

Assessing the future of encyclopedic content, Wales once argued that
"*Encyclopedia Britannica* will be crushed out of existence within five years." He
no longer believes that's the case, but Wikipedia's dynamic low-cost production system certainly makes it hard for *Britannica* to compete. "How can they
compete? Our cost model is just better than theirs," he says.[10]

It's hard to argue with him. Wikipedia is a fast, fluid, and comparatively inexpensive production model. Switching costs often hinder new
products or competitors, but not Wikipedia. Wales offers a zero-cost alternative that provides access to an unprecedented amount of information
with one click. Go to *Britannica* and you're looking at a laborious subscription process and $11.95 a month before you can get to the best content.

When you talk to Wales you understand that Wikipedia is merely the beginning. It's just the opening act in what promises to be a sustained and riveting journey. Wales is already exploring new frontiers. Dictionaries, textbooks, news, free-content libraries—the sky is the limit for Wales. He even muses about publishing spin-off books like the *The Wikipedia Guide to Rock and Roll* or *Wikipedia's History of World War II*, books ultimately written by the users, and profits that could be rolled back into the community to make it even better.

Where else could peer produced content take us? It appears it will take us very far indeed. As we explore in later chapters, people are self-organizing to produce music, news, TV, video games, and a variety of other forms of information and entertainment. Indeed, a growing number of entrepreneurs think that wikis can be harnessed for profit. In January 2005, Jack Herrick started wikiHow, an advertising-supported how-to guide built in the same open fashion as Wikipedia. Visitors can find free articles about everything from "how to get a mortgage" to "how to excel in high school"—all of them supplied by knowledgeable volunteer contributors who, in some cases, hope that they will soon share in the revenues the Web site generates. Sites like ShopWiki and Wikitravel are following in Herrick's footsteps. Wikis have unleashed a powerful force: a self-fulfilling, virtuous circle of cocreation that hierarchical models are powerless to stop or replicate.

IBM AND THE OPEN SOURCE EXPERIMENT

When Linus Torvalds first posted a fledgling version of Linux on an obscure software bulletin board, no one—apart from the most die-hard open source evangelists—would dare have predicted that open source software would be much more than a short-lived hacker experiment. And yet, within a few short years Linux has spawned a multibillion-dollar ecosystem and upset the balance of power in the software industry. Companies that once vied to control the lucrative computer operating system software market with their own proprietary solutions are now suddenly facing serious competition from a free alternative built by a loose network of programmers who aren't even out to make a profit.

As Linux rapidly gains ground, the industry is realizing that open source is a force to reckon with. Smart firms are learning how to coexist with and profit from the arrival of a new mode of software production.

And if there is one company that exemplifies this potential—along with the deep, wrenching transformation it entails—it's IBM, whose early foray into open source provides lessons for anyone seeking to harness peer production in their business.

IBM was an unlikely candidate to become a champion of peer production and a leader of the open world. After all, we're talking about Big Blue—the company that became huge by building and selling proprietary everything. For decades it created software that only worked on IBM computers. Tough luck if you wanted to port it to another vendor's hardware. IBM called it "account control." Detractors called it "hotel proprietary." That is, you can check out any time you like but you can never leave. But in a stunning reversal of strategy (and fortune) IBM has embraced open source at the core of its business in a way that few organizations of its size and maturity have dared.

It's fair to say that IBM did not set foot on this journey from a position of strength. Many of its proprietary offerings in Web servers and operating systems were failing, and the company was having a hard time unseating entrenched rivals like Microsoft. In an unorthodox move, IBM started investigating open source software, eventually donating large volumes of proprietary software code and establishing teams to help the Apache (Web server) and Linux (operating systems) open source communities.

Today Linux services and hardware represent billions of dollars in revenue, and IBM estimates it saves nearly a billion dollars per year over what it would cost to develop a Linuxlike operating system on its own. More than that, supporting open source has enabled IBM to undercut competitors such as Sun and Microsoft, who charge for operating system software, essentially commoditizing their offerings.

With IBM now years ahead of its competitors, the company's involvement with open source communities provides the quintessential example of how smart companies can harness self-organizing webs of independent contributors to create unrivaled value.

The Apache Experiment

When Linux emerged from the hacker fringes of the Internet, IBM, for decades the king of proprietary, homegrown operating system offerings, was in no mood to develop yet another new OS. It was an expensive and risky proposition with no guarantee of market adoption. Yet Linux could prove competitive to Sun or Microsoft, which made it attractive to IBM. By 1998 IBM was vigorously researching Linux and open source software generally. "At the time we had great concerns," recalls IBM strategist Joel Cawley. "Would they reject us? Will there be hostility to IBM? Will we face new legal issues that affect our ability to develop software?"

IBM decided to get involved in open source, but not initially in Linux. The company joined the Apache group, a team of programmers who had developed server software for Web sites. Apache already had about half the Web server market, and IBM's own product, Domino, had less than 1 percent. IBM did not have much to lose. In March 1998, its representatives met with Brian Behlendorf, head of the loosely organized group of programmers who maintained updates to Apache.

Both sides were a little wary. The free-software programmers were afraid of being tainted by IBM; Big Blue had legal and technical concerns about working with an ad hoc, globally distributed project team. Ultimately they struck a deal in which IBM committed to join the Apache community, release its code under open source rules, and participate much like any other contributor. IBM made modest financial contributions to set up the Apache Software Foundation, a legal entity that housed IBM's contract.

They worked fast, and on June 22, 1998, just three months after their initial meeting, IBM announced it would support the Apache server on all of its products. IBM put Apache into its WebSphere offering, and it took off. The deal was a watershed moment in the history of open source.

Joining Linux

The success that IBM experienced working with open source programmers and integrating Apache into its offerings readied the firm for its foray into Linux: By December 1998, the company was formally considering Linux strategies. IBM knew that Linux adoption was growing quickly. Customers were increasingly asking about running Linux on IBM hardware, and the company was finding that new hires from universities were fluent in Linux and supported open source.

At the time, IBM faced a strategic challenge: It was pinned between low-end hardware vendors, particularly Dell, and operating system vendors Microsoft (Windows) and Sun (Solaris). Linux offered solutions. It was a scalable operating system that would work well on small servers and could grow to handle heavier tasks. Because it was free, customers could try it easily. These advantages would help shift the locus of differentiation from operating systems to services and solutions, IBM's sweet spot.

In 1999, IBM formed a Linux development group. Its director, Dan Frye, says the toughest job in the early days was figuring out the right way to join the community. Linux consists of more than one hundred umbrella software projects, each with varying numbers of subprojects. Perhaps a thousand people contribute to the kernel, the heart of the operating system. Other groups handle libraries, drivers, and other components. IBM needed to decide which Linux communities to join. It found, as others who join open source communities do, that the best way to gain acceptance is to take on the unglamorous tasks that need doing. IBM helped improve Linux reliability through code testing, defect management, writing documentation, and open sourcing its own code and tools.

"One of the things we learned early on," says Frye, "is that people participate in open source communities as individuals. You are not employee X of company Y. You are a lone human being. The company you work for doesn't impress the programmers in the community. And each of these communities is different, so every time you want to work on something new, you have to learn about that community in order to join it and be effective."

As it began working on various Linux projects, IBM paid attention to getting the culture and processes right. Open source software communities run on instantaneous, transparent back-and-forth communications and rapid product iterations. Conversations use instant messaging, e-mail— whatever is fast. By comparison, internal company communications, attentive to internal sensitivities, are frequently slow and measured. Frye says, "When we were responding slowly with canned answers we weren't fast enough or transparent enough. It was not a level of technical exchange that was attractive to Linux developers." Frye told his team: "I'm unplugging you from the network. You can only communicate about Linux through the Linux community." And from then on the team used the same bulletin boards and chat rooms as Linux developers.

IBM also ran into a profound difference between open source and traditional software design. Though the steps—design, development, testing, maintenance, etc.—remain the same, open source communities tend to spend far more time and attention on implementation, testing, and support, and relatively less on user requirements and design specifications. A proprietary project may take months of planning and internal sign-offs before a single line of code is written. Open source projects can be kicked off by an individual who writes part of a program and posts it online. New code or compilations of the program may be published daily, enabling a global community of users to test and fix the product continually. And, since the end product is free and anyone can change the code, the product remains "in development" long after it is "released."

Frye learned that it was up to IBM, as a newcomer to the open source b-web, to adapt. IBM adopted a more open, transparent stance on collaboration with external developers. IBM employees had to be as candid within the open source community as within their own firm. Using the community's preferred communications channels, even for communications within the team, helped make IBMers into community members. Adopting open source programming tools, even though they were often less sophisticated than IBM's, enabled better collaboration with programmers outside IBM.

IBM not only accepted open source software products and processes but also its philosophy, which is to spur quality and fast growth rather than just profits based on proprietary ownership of intellectual property. At one time IBM could conceivably have released its own version of Linux—called a "distribution," in Linux parlance. However, it chose not to distribute the software and instead supports distributions from companies like Red Hat and Suse.

Giving up so much control is unconventional to say the least, but the rewards for doing so have been handsome. IBM spends about $100 million per year on general Linux development. If the Linux community puts in $1 billion of effort, and even half of that is useful to IBM customers, the company gets $500 million of software development for an investment of $100 million. "Linux gives us a viable platform uniquely tailored to our needs for twenty percent of the cost of a proprietary OS," says Cawley.

By most measures, IBM's involvement in the Linux community has been a huge win for both parties. At a time when reliability and trust were the big question marks surrounding Linux, IBM indemnified client risk. And Big Blue's early buy-in and financial commitment has greatly improved its competitive position vis-à-vis competitors like Sun and Microsoft. First, IBM gained a viable alternative to the Windows server on Intel-based platforms. Linux has also taken a bite out of Sun, depressing its profits and market share, and threatening its hardware business model.

Just as important, IBM has gained experience and knowledge in a vital new model of value creation. A company that was proprietary, insular, and vertically integrated fifteen years ago now partners extensively with the open source community and is considered a positive force for collaboration and openness. IBM enjoys the goodwill of thousands of independent and corporate developers who are committed to the Linux vision and community growth. Its partnering and collaboration skills and its specific knowledge of how to manage relations with communities it does not directly control are strategic tools competitors have yet to master.

Embracing Open Source Culture and Strategy

Open source has enabled IBM to speed innovation and off-load tremendous costs. From a strategic perspective, this approach to peer production is a form of collaborative outsourcing. And collaborative outsourcing works best in areas that are not core to your product or central to your business model. A number of lessons arise from these strategic considerations that may help you and your business.

First, play to your weaknesses. Look for areas in which your market efforts have floundered. Collaboration will be cheaper. IBM had already failed with Web servers and operating systems (OS/2), so it had little market share to lose by embracing open source. At the same time, look for opportunities that attract customers or have industry disruption potential.

Second, take a balanced approach. Ask yourself: Are you ready to go at it alone as leader of a peer production community? Or can you accomplish your goals by joining existing "movements"? In most cases, attaching yourself to existing movements that have momentum will create the best results. Do not abandon vertical integration and hierarchy. Instead, integrate proprietary and open source models. A blended approach allows you to adapt your strategies to the disruptive possibilities that arise as the project evolves—as IBM has done with "gifts" of free code to open source.

Third, adapt to community norms and clock speeds. Do not try to lead until you have built credibility in the community. And do not criticize. Dan Frye saw activities he did not like but kept quiet. Critiquing the community is a right reserved for those who have proved themselves by making valuable contributions. Use the community's preferred tools and communication methods. And remember: Online communities of all types tend to evolve faster than hierarchical processes. IBM had to make sure its engineers worked at the speed of the external community.

Finally, make it a priority. Firms considering open source or other self-organized communities in the production of intellectual property may be hypersensitive to the risks. When IBM was new to open source, perceived risks were high. "In the beginning we were looking at it at a very senior level," says Cawley. The company put the vice chairman in an oversight role, set up a Linux steering committee, and held monthly senior meetings to evaluate progress. "Over time," says Cawley, "we grew more comfortable and paid less attention. Now, open source is baked into the culture. It's part of the strategy kit bag."

Though IBM did not consciously invite this change at first, it has demonstrated remarkable adeptness in learning to use openness and self-organization as strategic weapons. The journey it has taken illustrates just how deeply open source is changing the very fabric and strategic orientation of the firm, and just how far the open source revolution will take information-based industries, with software as their beacon, down the path to exciting frontiers in innovation and value creation.

THE OPEN SOURCE ECOSYSTEM

IBM provides a surprising example of how a large, mature company with an engrained proprietary culture can embrace openness and self-organization as catalysts for reinvention. But to focus only on IBM and Linux would mean missing a newer and an arguably more important trend—the rise of a vast and vibrant ecosystem of start-up companies that are driving the next wave of development in open source business applications and services.

This new breed of software start-ups promises to redefine the very meaning and limits of open source. By offering low-cost, open source solutions companies like Digium, Medsphere, Pentaho, and SugarCRM are making inroads in the once impenetrable world of enterprise software—the software that large enterprises run to manage data, share knowledge, track projects, deploy resources, and generally make their businesses more efficient. Open source application vendors could be the force that brings affordable enterprise solutions to the masses of businesses that could never afford an Oracle database or an enterprise resource planning (ERP) system from SAP. And who knows, they just might empower a whole new revolution in business productivity, and perhaps even kick-start a renaissance for the small and medium-size business.

Is it possible that all software can be peer produced as opposed to being created by companies? Linus Torvalds has changed his mind on this. "I'm lousy at making predictions," he says. "I used to think only the operating system could be done using open source, but I was proven wrong." He thought, for example, that "no one would ever want to self-organize to create a database, because it's too boring." He's now concluded that open source communities could create almost any software, except for small niches where it might be too difficult to bring a community together.[11]

If Linus's change of mind is valid, how might this unfold? And what would be the implications for the software industry—a dynamic and critical part of most economies? To understand this progression, think of the open source software movement as two oncoming waves, with roughly a decade between them. The first wave brought us the plumbing: open source Web servers, operating systems, and the various pieces of code needed to run the Internet. Techies call it the LAMP stack, short for Linux, Apache, MySQL, and Perl/PHP. Linux you now know about. The Apache Web server runs on more than one hundred million sites. MySQL's open source database application has been downloaded more than one hundred million times. Nearly three-fourths of all Web sites use the open source programming language PHP.

The first wave of open source provided the foundation for the really expensive and complex applications that enterprises use to run their businesses. But when it came to the enterprise applications themselves, open source hit a wall. Indeed, for almost as long as software has existed, these enterprise-proof applications have been the preserve of large software houses like SAP, Oracle, and Microsoft. Now that's changing, with a second wave of open source.

Today everything from customer relationship management (CRM) to enterprise resource planning (ERP) to content management and business intelligence—basically any enterprise software application you can think of—is becoming available in open source. The full array of new open source ventures is astonishing (around ten thousand and growing), but just consider two examples.

Pentaho provides open source business intelligence that competes with commercial applications provided by Cognos and Hyperion. Like commercial vendors, Pentaho's solution provides enterprise-class reporting, analysis, data mining, and workflow management capabilities that help organizations handle their data more effectively. While customers fork out large dollars for commercial BI software (with related consultancy on top), Pentaho's baseline software is free. Like other open source vendors, it generates revenue from support, training, and consulting to customize the software for a company's specific requirements.

Medsphere provides open source software for small and midsize hospitals, enabling them to manage electronic health records that track each patient's information from lab tests through to prescriptions. While there's already plenty of hospital-management software on the market, most small facilities can't afford it. And yet, 80 percent of the hospitals in the United States have fewer than three hundred beds. At a quarter of the cost of proprietary systems, with support and installation included, Medsphere's system is proving to be popular with these facilities.

It's too early for anyone to be claiming victory, but the cost advantages for these companies are considerable. They don't need to hire armies of salespeople or engineers because the open source community does a good deal of the heavy lifting by helping produce, debug, and evangelize the software. The money they do spend usually goes into developing value-added features and services, which is a big change from the proprietary enterprise model, where up to 70 percent of the costs funnel into sales and marketing.

Companies frustrated by the expense or hassle of proprietary software are increasingly willing to give open source a try. After all, they can usually download a trial version off the Web for free. If the open source upstarts manage to get a toehold, there will be profound consequences for incumbent software vendors whose business models rest on the whopping fees firms pay to license enterprise solutions.

Managing Open Source Complexity

When it comes to open source, complexity is both a feature and a bug. The three rules of open source—nobody owns it, everybody uses it, and anybody can improve it—may be the source of endless innovation, but they're equally a source of endless frustration for the poor IT managers who deal with the resulting complexity. The reality is that there are just too many choices. Businesses wishing to harness open source must first wade through a vast heap of applications. Which ones are quality? Which ones are not? If you can sort that out, congratulations, you've gotten past step one. Now, by some stroke of luck or magic, you'll need to make all of the disparate applications work in concert.

And still, that's just the beginning. A typical open source software program has a shelf life of days or weeks. Mainstream software houses, by contrast, can take years to turn around new versions. This rapid, iterative development style means the software is constantly improving. But it also means that businesses seeking continuity and reliability may find themselves at the mercy of large, and largely anonymous, communities of programmers that pursue advances at a whim.

All of this complexity pleases the big software houses. They point to the costs and risks of going open source and flaunt their comparatively stable solutions as the responsible mainstream choice. And yet, if the compatibility, integration, and support issues could be cracked then the proprietary vendors could be in trouble. Few have assumed such a complex task is possible without the resources and command structure of a large, vertically integrated firm. But then, the big software houses didn't exactly count on Kim Polese stepping up to the plate either.

Kim Polese is as close as it comes to being a high-tech rock star. She made a name for herself as chief executive officer and cofounder of Marimba, a leading provider of Internet infrastructure management solutions. Before that, she spent just under a decade at Sun Microsystems, where she was a key architect of Java (Sun's flagship programming language). Her vision, wit, and enthusiasm made her a luminary in the first Internet boom. Her good looks didn't hurt either, as magazines such as *Wired* clamored to get her photo on their covers. But beneath all the pizzazz is a person of strong vision, business acumen, and deep technical ability.

One evening in the summer of 2004 Polese was invited to dinner with Google's Eric Schmidt and Ray Lane, a software veteran and former president of Oracle. Ray was now a general partner at Kleiner Perkins Caufield & Byers, a prominent venture capital firm in Silicon Valley. He had been incubating a company that was working on a solution to the open source complexity problem, but he needed a natural leader. Polese was intrigued, and together they hatched a plan. The name of the company was Spike-Source, and Polese was to become CEO.

Polese recalls walking away from the dinner feeling refreshed. "I just had that feeling like I did back in ninety-five, where if there was ever a time to start a company, you know, this was it," said Polese. "I was seeing incredible innovation that I'd never seen in the twenty years I've been in the software world. A lot of the problems that we had always experienced with software were suddenly becoming tractable with open source." Polese was on a mission: bring open source to the mass market by solving one of the thorniest software challenges, and by doing it in true open source fashion.

Integration and interoperability are age-old problems in software generally, and not just in open source. These problems exist in large part because there have always been silos and walls between various competing vendors. None of these proprietary vendors were willing to share interfaces or collaborate to properly test and integrate applications. This produced unpredictable, and sometimes disastrous, results when customers tried to fuse their applications.

The current wisdom is that industry consolidation is the answer. A recent spate of mergers and acquisitions in the commercial software world (most notably by Oracle) suggests that this process is under way. In the case of open source software, large proprietary vendors could opt to buy up smaller open source competitors, thereby subsuming the open source movement. Polese and plenty of others, however, see it differently, and are opting for a collaborative solution.

With the second wave of open source, true collaboration and integration between applications is a real possibility. "In fact, it's happening naturally," says Polese, "in a totally organic way because the values of collaboration are engrained in the open source way of doing things. All these different open source projects and companies are beginning to work together, and that's what's needed to make software dependable and useful."

Polese says open standards and the collaborative infrastructure of the Web allows multiple companies and communities to work together to resolve the integration, testing, and support issues. "Not only can we now run tens of thousands of tests in an automated fashion, we can enable thousands of people across the community to contribute to a giant knowledge base of what works with what," she says.

Now her company, SpikeSource, has a breakthrough innovation. They call it an "automated test framework" that crunches over thirty thousand different test runs nightly across hundreds of components, six operating systems, and six language run times, and then cranks out an integrated solution (or stacks, as they're called in geek speak). Think of it as a massive digital assembly line that takes an assortment of different bits and pieces, shakes them up, and slots them together to build a unified and well-oiled machine. Whenever a new open source application or update comes through on a bulletin board, SpikeSource will test it and integrate it into the stack. Downloading the stack is free. SpikeSource makes its money providing customer service and support.

Not every Fortune 500 company is convinced that open source is ready for prime time. So, in typical disruptive fashion, SpikeSource and other open source application providers are targeting small and medium-size businesses, a market that incumbent vendors are generally unmotivated and ill-equipped to enter. The low-cost open source model enables them to make their service offerings affordable for many organizations and for uses that weren't possible before. This in turn is driving a wave of democratization in software.

"These mid-sized businesses or small businesses are eagerly embracing open source, not because they're open source zealots," says Polese, "but because they can get incredible functionality at a fraction of the cost." Suddenly enterprise-hardened tools for managing sales teams, customers, content, data, and resources are within reach for legions of small and medium-size businesses. Now they have a chance to reach or exceed the levels of management efficiency and effectiveness only seen in large firms.

The Future of Open Source

Open source, it seems, may at last be coming of age. But as much as it represents an exciting new frontier for start-up businesses, it raises tough issues for the open source community at large. In the first wave of open source, early movers such as Red Hat and IBM coexisted quite comfortably with the loose alliance of programmers that collaborate on open source software. Key to this has been their willingness to respect community norms, adopt open source processes, and strike a healthy balance between what they receive and what they give back to the community.

Hundreds of millions of dollars of VC money are now pouring into open source start-ups. With heightened pressure to produce healthy margins, it's unclear whether this new generation will perform the same deft balancing act that has enabled for-profit and nonprofit communities to mix amicably. Open source purists worry that a growing tide of for-profit ventures could extinguish the ethics of sharing, reciprocity, and openness that lie at the heart of the open source community's value system. Indeed, firms face pressures on both sides: They will need to give away enough valuable code to satisfy collaborators, while still holding back something important enough that customers will want to pay.

And yet, new models of cocreation and collaboration between firms indicate that this trade-off may be illusory. In fact, the days where collaborating in the open source community meant pulling an all-nighter after a long day of mind-numbing programming may be coming to an end. The opportunities have never been greater for open source loyalists to directly monetize their contributions. They can form their own businesses and still collaborate across firm boundaries in the same old open source fashion.

It starts with the way that open source companies think about software and the business models for delivering it. Collaboration is not an afterthought; collaboration is designed into the software from scratch. "In the open source world," says Polese, "every discrete component is made to be part of a larger ecosystem. When developers sit down to create a component or a project, they start by thinking about how it will interoperate with all of the other pieces out there. So it's a very different approach to writing software and to making a business in the software world."

It's very different indeed. In the past, the software business was about capturing the customer, locking them into your platform, and locking out the competition, essentially creating better furnished jail cells for the customer—welcome to the hotel proprietary.

"Open source is about knocking down those walls," says Polese, "and about actively seeking ways from the very beginning to make your software work better with everybody else's software. This is opening up a huge new wave of innovation, and software is getting much better at an accelerated rate. There are more people pounding on it, more people using it, more people participating in creating it, and more emphasis on collaboration and integration."

As for making money, well, it's all about adding value. Polese says, "You keep your customers happy by offering better maintenance and support. You offer interoperability with other software applications. And you're always adding on, you're making it better and better over time."

Can proprietary software vendors see the writing on the wall? We sure hope so. Proprietary software spells rapid obsolescence if these companies don't find a way to coexist with the peer producers.

WHY THE OPEN SOURCE CRITICS ARE WRONG ABOUT FREE ENTERPRISE AND PROFIT

In *The World Is Flat*, an otherwise helpful book, Thomas Friedman argues that open sourcing is a powerful flattener but concludes in the same breath that in a world of open source it will no longer be clear who owns what or how individuals and companies will profit from their creations. He seems influenced by open source detractors who argue that this is latter-day socialism and an attack on free enterprise and the right to make a profit.

Like many critics, Friedman is not seeing the forest for the trees. He sees free software, but not the multibillion-dollar ecosystem that surrounds open source. He sees free encyclopedias, but not the rich cultural and educational opportunities that envelope a living, breathing, dynamic repository of knowledge updated by a vast self-organizing community. He sees the potential for Chinese firms to rip off the designs of American manufacturers, but not the opportunity for BMW to invite its customers to codesign the telematic features for future cars.

Skeptics make the same mistake when looking at projects like the

Apache Web server. "Gee, free server software that powers seventy percent of all Web sites—one of the most successful business software projects ever," they say. "But Brian Behlendorf, the guy who orchestrated the whole thing, didn't make a dime off it!"

Behlendorf never intended to make money off Apache, mind you. He and his fellow developer buddies just wanted a better Web server. Nevertheless, open source skeptics reason that if he *had* sold Apache for the same price as Microsoft's comparable server software, he might have earned sales upward of a billion dollars.

But then, if Apache had cost the same as a server from Sun or Microsoft it's unlikely that it would have outrun these companies in the server market. And all of the innumerable value that is created by and around the Apache ecosystem might not have materialized, including the legion of small and medium-size enterprises that are thriving on this low-cost infrastructure.

Behlendorf, now a veritable legend in the open source community, has since gone on to found CollabNet, a successful company that provides workflow and collaboration tools required to collaborate across firm boundaries and integrate self-organizing developer communities into their rigorously structured product-development processes. The company now has a growing list of Fortune 100 clients that are eager to tap the efficiency benefits. Had Behlendorf not gained worldwide notoriety from the Apache project, it's unlikely he'd be where he is today.

Embracing open source means embracing new mental models and new ways of conceptualizing value creation. It has long been fashionable to say that public goods are inimical to wealth creation. Economists and business leaders have frequently argued that what goes into the commons takes away from the mouths of private enterprise. Of course, a growing number now realize that this is nonsense. Without the commons there could be no private enterprise. As Linus Torvalds aptly put it, "That's like saying that public roadworks take away from the private commercial sector." Even if public ownership of key aspects of the transportation network forecloses opportunities for private profit, the gains to the rest of the economy make these losses look minuscule.

For Torvalds, Linux is like a utility. It provides the basic infrastructure on which software developers can build applications and businesses. "It allows commercial entities to compete in areas that they really can add value to, and at the same time, they can take all the 'basic stuff' for granted," he says.

"This is especially important in software," he continues, "where proprietary source-code at the infrastructure level can actually make it much harder for other players to enter the market. So if anything, open source is what makes capitalism in software possible at all. Without open source, you'd have just a set of monopolies: effectively, economic feudalism." In fact, he finds it rather ironic that those favoring proprietary software would attack Linux as unfair. "At minimum they should accept it as fair competition. We don't have proprietary lock-in, financial capital, government subsidies, distribution systems, or other advantages of private companies," he says. "This is not socialism; it's the opposite—it's free enterprise."

The changing nature of the Linux community itself provides further evidence of this. No longer just an ad-hoc collection of individual volunteers, many of the participants in the Linux ecosystem are paid employees of Fortune 100 tech firms. This really ups the ante on peer production and foreshadows some big structural changes in the economy. Indeed, as companies assign serious resources to peer production communities, questions about when to contribute and how to harness the commons are arriving at the heart of corporate strategy. It's no longer just about boundary decisions (what's in, what's out), but big questions about where to play, when to cooperate on shared infrastructures, and when to differentiate and compete.

For IBM's Joel Cawley, contributing to shared infrastructures like Linux does not decrease opportunities to create differentiated value; it increases them. It's just a matter of thinking about value creation differently. "One of the things you can get confused about in doing strategy," says Cawley, "is losing sight of where real value comes from. If you are constantly creating new value then you have opportunities to harvest that value." In other words, shared infrastructures that grow and evolve constantly force firms that contribute to them to grow and evolve constantly too. And so long as they add value, there will always be healthy profits.

With that promise comes a warning. "At some point you may stop creating new value," says Cawley, "and when you do, if you continue to harvest, then at some point you're no longer earning the harvest and, in fact, you may not even be doing yourself much good, because you're no longer off creating new value; you're stuck in a rut."

Profiting from peer production communities like Linux may never be quite as direct or straightforward as profiting from more conventional products and services. It's a new skill that requires companies to recognize and seize opportunities to build new products and services on top of vibrant open ecosystems—ecosystems where new value is always being created for a variety of ends and motivations.

Companies need unique capabilities to work in these environments. To leverage the benefits faster and more effectively than competitors, for example, companies need capabilities to develop relationships, sense important developments, add new value, and turn nascent knowledge into compelling customer value propositions. Joel Cawley calls it "the ongoing process of regeneration, of creating new sources of value. And that is what a lively functioning enterprise is all about."

Understanding and applying this new approach to competitiveness means dispelling some deeply rooted biases. The conventional wisdom says that sharing IP and other resources creates a public good where everyone automatically shares in the benefits, and there is no way to generate private returns. On the contrary, our research points to a multitude of ways in which smart firms can harness peer production to drive innovation and wealth creation. Here's a selection of some of the key benefits of peer production for business.

The Key Benefits of Peer Production for Businesses

Harnessing external talent. Today the speed and complexity of change is such that no one company can create all the innovations needed to compete in information technology, or in any other industry. Science and technology are advancing fast and individuals and companies are using and deploying new knowledge in unanticipated ways. Smart firms can harness this innovation by using peer production to involve way more people and partners in developing customer solutions than they could ever hope to marshal internally.

Keeping up with users. Not since the elimination of Netscape has Microsoft experienced any real competition in the Web browser market. Now Netscape has been reborn as Mozilla Firefox, an open source Web browser that allows users to alter the code and create plug-ins and customized "extensions" that can then be downloaded by any user. Over the last few years Firefox has crept up on Microsoft to claim a 28 percent share of the market. This shows that if you do not stay current with users, they invent around you, creating opportunities for competitors.

Boosting demand for complementary offerings. Participating in peer production communities can boost demand for complementary offerings and provide new opportunities to create added value. Firms that engage with the open source community, for example, generate returns from increased services, support, and hardware sales, and this in turn opens up an opportunity to create more IP, just as the growing popularity of Wikipedia has convinced Jimmy Wales that there may be a market for a Wikipedia-branded line of books.

Reducing costs. By collaborating with open source communities, companies can reduce costs dramatically. IBM estimates it has saved $900 million per year compared to what it would have to spend on creating and maintaining an operating system in-house. Companies must dedicate resources to filtering and aggregating peer contributions. But these types of collaborations can produce more robust, user-defined, fault-tolerant products in less time and for less expense than the conventional closed approach.

Shifting the locus of competition. Publishing intellectual property in non-core areas that are core to a competitor can undermine your rival's ability to monopolize a resource that you depend on. Many pharmaceutical companies, for example, contributed to the public human genome project because their business model depends not on patenting genes but on discovering new drugs. In the software industry, publishing code has enabled IBM and Red Hat to migrate the locus of competition from operating systems to applications, integration, and services.

Taking the friction out of collaboration. As the need for collaboration increases, firms are finding that problems related to ownership and exploitation of intellectual property can make proprietary collaborations difficult.

Participants may have trouble clearly defining the boundaries of their intellectual contributions, and concerns about public disclosure of proprietary information and disputes over future patent rights can create friction. Avoiding these problems is one reason why a growing number of firms are embracing open models of collaborative innovation.

Developing social capital. When firms join a peer production community sharing is the continued price of admission to the community from which the firm derives various benefits. This is why firms like IBM, Sun, Nokia, and others are granting open source communities royalty-free access to some of their patents. In exchange, they obtain a "license to operate" in the community—a form of tacit permission to harvest some of the value created in collaboration with community members.

Collectively, these examples suggest a range of ways in which peer production creates value and competitive advantage. Some of these benefits accrue to all comers and others boost the competitiveness of firms positioned to leverage them. Peer production has its share of trials and limitations, however. It means less control and requires practitioners to learn and abide by the rules of scientific and creative communities. It means devising new incentive structures and forging creative business models that allow firms to simultaneously harvest and contribute. And it means investing in infrastructures for collaboration while carefully considering issues such as IP diligence and indemnity. Companies that wish to reap the benefits must be equally prepared to meet these challenges. In the concluding chapter of this book we will take some time to address these and other challenges more thoroughly.

PEER PRODUCTION IS HERE TO STAY

Peer production communities are taking their place alongside open markets and hierarchical firms as an important alternative competitive strategy and way of organizing work. They leverage basic human motivations to turn work that once would have been considered nonremunerative into substantial economic value. Peer production will continue to grow in importance because key enabling conditions are present and growing. These include: access to computing power and applications; transparency; globalization; the democratization of knowledge and skills; and the increasing complexity of systems.

All firms need to take stock of how self-organized innovation and production might be introduced into their industries and what the economic impacts will be. They need a new strategy agenda that includes, at a minimum, an ongoing analysis and audit of the potential vulnerability of current models to self-organizing competitors, and an examination of opportunities for the firm to leverage peer production communities to lower costs, drive innovation, and gain competitive advantages.

The greatest risk is not that peer production communities will undermine an existing business model, but that a firm will prove unable to respond to the threat in time. Firms must invest in the technology and business architecture to become truly networked and open enterprises, and engage in collaborative networks that help build the cultural and strategic capabilities to leverage peer production. They should heed the words of Joel Cawley, who looks back on IBM's nearly ten-year experience with open source software and says, "Having gone through the journey, we are comfortable with open source and all the things connected to it. We see open source as part of the kit bag of strategy now. We understand that if you don't do it your competitors will. And then where will you be?"

4. IDEAGORAS

Marketplaces for Ideas, Innovations, and Uniquely Qualified Minds

The late-nineteenth-century chemist and microbiologist Louis Pasteur famously said that chance favors the prepared mind. The same could be said of innovation. Companies face tough dilemmas every day for which there is a uniquely prepared mind somewhere in the world who possesses the right combination of expertise and experience to solve that problem. The trouble is that uniquely prepared minds can be illusive, like the proverbial needle in the haystack. But today, a new marketplace for ideas, innovations, and uniquely qualified minds is changing everything, and Werner Mueller is a perfect example of what's happening.

Werner Mueller is a brilliant chemist who worked most of his career at Hoechst Celanese. Mueller had always been passionate about doing science. But every time he did a good piece of work he got promoted, and as a result he got to do less and less science.

When Mueller retired he built facilities in his home to do the two things in life he loved the most, chemistry and woodworking. He spends half a day on each passion. Then one day he stumbled across a Web site called InnoCentive that listed a variety of scientific challenges that needed answers. What's more, there were some handsome cash rewards for anyone who could provide a workable solution. Mueller thought this was great. "I've now got a set of challenges that I can work on," he said.

One day, late in 2001, a pharmaceutical company required early stage raw material for a product it was bringing to market. Though the chemical compound wasn't enormously expensive, the method for producing it was inefficient, and this contributed a great deal to the final drug cost. The internal R&D team was struggling to find a solution, and the project was already over budget. So the team posted the problem on InnoCentive, where

Mueller soon spotted it. It was a problem that Mueller recognized from his decades of experience as a chemist. He went to work in his lab, and not long after he submitted a valuable solution. The company was delighted—it was a solution it had not yet considered—and Mueller was $25,000 better off. He reinvested the award into his lab, and turned his retirement hobby into a full-fledged consulting business.

Mueller's case is hardly isolated. He is one of two hundred thousand scientists from 175 countries who have registered with InnoCentive to provide solutions to companies such as Boeing, Dow, DuPont, Novartis, and Procter & Gamble. Launched by U.S. pharmaceutical giant Eli Lilly as an e-business venture in 2001, a number of Fortune 500 companies now tap InnoCentive to extend their problem-solving capacity. This visionary matchmaking system links experts to unsolved R&D problems, allowing these companies to tap the talents of a global scientific community without having to employ everybody full-time.[1]

Werner Mueller and the story of InnoCentive points to a deep change in the way companies innovate. Companies can tap emerging global marketplaces to find uniquely qualified minds and discover and develop new products and services faster and much more efficiently than they have in the past. We call these marketplaces Ideagoras, much like the bustling agoras that sprung up in the heart of ancient Athens. In those days, agoras were the center of politics and commerce for the burgeoning Athenian citizenry.[2] Modern-day ideagoras such as InnoCentive serve a more specific purpose: They make ideas, inventions, and scientific expertise around the planet accessible to innovation-hungry companies.

Tapping an ideagora is a bit like having an eBay for innovation. Just as millions of people connect on eBay every day to auction off Beanie Babies, motorcycles, computers, and used suits, companies can procure just about anything they require to launch a product in an online marketplace that matches buyers and sellers of innovation. Ideas, innovations, sources of financing, and millions of enterprising individuals now thrive outside traditional enterprises. Ideagoras can help companies get hold of them.

InnoCentive works a little bit like eBay. Companies—or "seekers"—anonymously post R&D problems on the InnoCentive Web site, while "solvers" submit their solutions in a bid to capture cash prizes ranging from

$5,000 to $100,000. InnoCentive chairman Darren Carroll says, "We're breaking down traditional laboratory doors and opening up an exciting new frontier where solution seekers—well-respected global corporations—can reach beyond their traditional R&D facilities and tap into more of the brightest scientific minds in the world."

The promise for seekers is that a large, diverse network of talent will solve well-defined problems faster and more efficiently than an internal R&D group. Indeed, InnoCentive solvers have been known to yield surprising results that lead companies down paths they would never have considered. CEO Alf Bingham says it's a property of making the problem accessible to many, many minds: "It produces a diversity of thought about the problem that can often make the solution rather unique."

InnoCentive is just one example of a growing number of real businesses on the verge of exploding into vibrant virtual marketplaces. Nine-Sigma, InnovationXchange Network, Eureka Medical, YourEncore, and Innovation Relay Centers offer similar services. The aptly named Your-Encore, for example, works much like InnoCentive, except it explicitly recruits retired scientists like Werner Mueller who are looking for some new challenges to pass the time.

Though today's nascent ideagoras have yet to reach truly eBay-like proportions, think of them as the first virtual trading floors in an emerging global idea bazaar. As more companies embrace the principles of wikinomics—openness, peering, sharing, and acting globally—these ideagoras will come to fruition, fueling an increasingly active trade in technology, intellectual capital, and other key innovation ingredients.

Tapping this eBay for innovation would forever change the R&D process and deliver significant economic efficiencies. Companies would no longer invent first and ask questions later. They would ask: "What do our customers really need?" and then scour the "eBay for innovation" for the necessary inventions and technologies. R&D labs would be ambidextrous: building on core capabilities internally, while acquiring the greatest, most complementary ideas externally. The deep-rooted "plan and push" modality would give way to a new approach to innovation where companies engage and cocreate with the best available talent, wherever it is in the world.

The old notion that you have to motivate, develop, and retain *all* of your best people internally would be null. Of course, you'll still need great internal talent. But increasingly, you should assume the best people reside outside your corporate walls. With an eBay for innovation, however, a massive reservoir of talent would be a few clicks away, at most.

This eBay for innovation could set the stage for an exciting new division of labor. Some organizations will excel in invention. Others will excel in marketing. Those organizations that can't successfully market all their good ideas could easily sell or exchange them. Firms with no comparative advantage in invention can access leading-edge technologies for much less than it would cost to develop in-house.

With the right approach, firms and industries could cross fertilize. One industry's technologies may create unanticipated efficiencies in another industry. It turns out, for example, that the same material that gives diapers their incredible absorption capacity is also useful for the undersea cables that string telecommunications lines between continents. So the diaper manufacturer licenses its technology to the cable manufacturer, gaining a new revenue stream and a better return on its R&D dollars. The cable manufacturer and the telecommunications companies gain a welcome innovation, one that they might never have conceived of internally.

The new ideagoras offer significant opportunity to small and medium-size firms too. Those without the management or marketing muscle to take on large firms can still earn healthy returns by selling their world-class ideas and technologies to others who commercialize them. And in the same way that markets such as eBay or Google Adwords have spawned a rich ecosystem of codependent firms, companies that become efficient at operating on the InnoCentive platform (as one example) might increasingly choose to focus purely on problem solving and intellectual property creation. Some contract research organizations are already doing this.

Meanwhile, a growing number of engineers and scientists could become free agents if they choose—setting up their own businesses and participating in a global marketplace for uniquely qualified minds. About one million of the most active traders on eBay have quit their day jobs and now make their living selling new and used goods full-time. So why can't a retired corporate scientist or an aspiring Ph.D. student use ideagoras as an outlet for their great ideas and an opportunity to solve problems for some of the world's most advanced companies?

All considered, ideagoras could lower transaction costs, deliver innovation faster, and make all participants in the marketplace more efficient. Customers would get more of what they want for lower prices. All of this is within reach. Companies just need to shake off the old approach to innovation and embrace the new, and that's where we're going now.

THE WORLD IS YOUR R&D DEPARTMENT

The reigning orthodoxy in innovation has long been that it is best to create and commercialize ideas within the confines of closed entities. Indeed, for most of the twentieth century this is precisely how innovation proceeded. Companies worked internally to turn the latest scientific breakthroughs into products the market wanted. They rarely looked outside for new ideas or inventions. One might even argue that they didn't need to.

After all, the important advances in technology were already happening inside large, well-funded industrial R&D machines such as Bell Labs and Xerox PARC. The labs of firms like DuPont and Merck attracted the most talented Ph.D. graduates from the leading universities from which they harnessed the revolutionary developments in chemistry and biology to pump out life-changing products and medicines.

Today's landscape is different—very different. Industrial economy knowledge monopolies are breaking down rapidly. The means of creation are open and proliferating. Innovations that once germinated in the R&D labs of large firms now flourish in a variety of settings. G8 countries can no longer expect to monopolize advanced scientific research. Yesterday's corporate leaders can no longer dominate their fields or dictate the pace of development.

Science and technology now evolve at such a great speed that even the largest companies can no longer research all the fundamental disciplines that contribute to their products. Nor can they can control an end-to-end production process or seek to retain the most talented people inside their boundaries.

Most managers will admit that they are currently nowhere near tapping the potential that ideagoras offer. An estimated 90 percent of research and development is still performed internally. "Most R&D organizations cling to the invention model," says Larry Huston, P&G's vice president for innovation and knowledge. "And in the invention model you develop bricks-

and-mortar infrastructure, you recruit the best people, and you develop global presence. Once you're global, you start to do R&D in different parts of the world, and then the next stage is to link them together so you can transfer ideas internally." The problem with these incremental changes is that they are "bandages on a broken model," says Huston. Something much bigger and bolder is required to master today's realities.

Companies should fundamentally rethink the invention model, and erect a new model based on the fluid exchange of ideas and human capital. "The question today," says Huston, "is how do you create a vibrant connections marketplace where you leverage other people's talents, ideas, and assets quickly and move on. Alliances and joint ventures don't open up the spirit of capitalism within the company. They're vestiges of the central-planning approach when instead you need free market mechanisms."

Huston's right. Acquisitions, alliances, joint ventures, and selective outsourcing are simply too rigid, and not scalable enough, to drive growth and innovation at a level that will make companies truly competitive. Smart companies will treat the world as their R&D department and use ideagoras to seek out ideas, innovations, and uniquely qualified minds on a global basis. As P&G CEO A. G. Lafley put it, "Someone outside your organization today knows how to answer your specific question, solve your specific problem, or take advantage of your current opportunity better than you do. You need to find them, and find a way to work collaboratively and productively with them." That's what ideagoras are for.

Solutions in Search of Questions

Ideagoras come in two principal flavors: solutions in search of questions and questions in need of solutions. Think of a classifieds site like craigslist, except rather than job ads and personals it posts a list of ideas and inventions that are "for sale" or "wanted."

Solutions in search of questions are those 70 to 90 percent of ideas and inventions that go unutilized; inventions that companies pursue for investigational purposes and yet never quite make it off the runway. For one reason or another they end up on the shelf, often because they are too costly or a poor fit with a company's brands and strategy. In other cases companies have great technologies that they are leveraging in their core markets. But the technologies have promising applications in other markets or industries that they are unprepared to enter.

Questions in need of solutions are just that: unanswered problems, queries, or uncertainties that—for reasons related to cost, timing, or lack of expertise—have not been addressed internally. We'll talk about these ideagoras in the next section.

The online technology transfer marketplace yet2.com started out as the former type of ideagora: a place where companies could post underutilized assets they were seeking to license externally. It was founded in 1999 as one of the first marketplaces of its kind. CEO Phil Stern explains that the idea arose as early supporters such as Boeing, DuPont, Honeywell, and Procter & Gamble all reported the same dilemma: They were sitting on mountains of intellectual property that they weren't using internally and looking for ways of monetizing it.

For Procter & Gamble the prospect of listing underutilized assets with yet2.com presented a potential windfall. The consumer products giant is the proud owner of over 27,000 U.S. patents. It would be even prouder, however, if more of these patents were fattening the bottom line instead of its legal costs.

In the late 1990s P&G launched an internal survey and discovered it was spending $1.5 billion on R&D, generating lots of patents, but using less than 10 percent of them in its own products. Yes, that's less than 10 percent! It's a figure that few companies admit to publicly, but Stern figures that this troubling statistic is broadly representative of the state of play in many research-intensive industries.

When the survey results hit Lafley's desk, they triggered a dramatic change in philosophy. Lafley saw it as a wake-up call and led the charge to open up the patent portfolio. The company, once renowned for its insularity, now sees external marketplaces for technology as one key pillar of its innovation strategy. P&G now makes every patent in its portfolio available for license to any outsider, as long as it has existed for at least five years or has been in use in a P&G product for at least three years. This keeps everybody on their toes and encourages employees to constantly replenish the company's stock of IP. And by sending the licensing revenue back to the business units, the strategy provides additional incentives for innovation.

The problem for P&G and other companies in its position was that finding applications and buyers for these innovative technologies could be highly inefficient. Until quite recently, the ability to move promising inventions in or out of the firm depended almost exclusively upon the personal networks of high-level executives and close interaction among small clubs of major innovators. In most cases, firms seeking to buy or sell new inventions and technologies would call up close associates in their industry.

While sophisticated patent searches aided the process of identifying desirable technologies, they typically produce more dead ends than leads. After all, patents are shrouded in lawyer-speak and say nothing of the willingness of the patent holder to share or license their technology.

Online exchanges, by contrast, promised to improve liquidity by expanding the universe of opportunities. They could also reduce search costs by easing the process of matching buyers and sellers. It's a promise that marketplaces like yet2.com are now on their way to delivering.

By visiting yet2.com companies can browse a list of billions worth of available technologies. The exchange's network of five hundred clients has access to roughly 40 percent of the world's R&D capacity. Indeed, with nearly one hundred thousand registered users, Phil Stern estimates that 80 to 90 percent of Fortune 500 companies have personnel actively listing and scouting new technologies.[3]

P&G is not alone in its affinity for IP licensing. Pioneers such as AT&T, IBM, and Texas Instruments have built their licensing practices into highly lucrative and scalable businesses. IBM in particular makes most of its IP available on a nonexclusive basis to partners and competitors equally. As compensation it rakes in over $1 billion in licensing revenue annually.

In fact, virtually all companies with sizable patent holdings are now busy mining their portfolios, looking for licensing-out opportunities, and taking technologies off the shelf that can bring in revenue. Using online marketplaces allows them to cast wider nets across multiple, and often unrelated, industries. Dave Christensen, a licensing executive at GE, explains that Internet-enabled technology transfer has boosted the global reach of its business relationships. "A lot of our licensing revenue now comes from companies in Asia," he says, "and the Internet makes our technology easily accessible to this large group of potential licensees."[4] Companies like GE can use ideagoras to buy and sell ideas and technologies on a much broader global playing field.

Another upside of marketplaces like yet2.com is that they allow smaller firms to participate in the economy in ways that were once impossible. While small companies often get out-muscled in product markets, ideagoras become a sort of alternative path to market where they can place their ideas or inventions up for auction. In a world where large firms are constantly scanning the landscape for new ideas and technologies, ideagoras will provide growing opportunities for small firms to become R&D suppliers.

Small firms are equally in a position to take advantage of the buy side of the equation. Large companies often place underutilized assets in marketplaces like yet2.com because the ultimate consumer market is too small to pursue. Niche players can take advantage of technologies that took tens of millions to develop. What's more, they spend no up-front capital to access them.

P&G, for example, recently used yet2.com to identify a buyer for a transdermal drug-delivery technology. The innovative system transfers large drug molecules like insulin through the skin so that a person with diabetes could wear a patch much like those used to help people quit smoking. P&G built a prototype but stopped short of commercializing the technology. Now a small company called Corium that specializes in drug delivery systems is set to launch the product. And, having now been introduced through yet2.com, the two companies are exploring further avenues for collaboration.

Questions in Need of Solutions

Most of yet2.com's clients started with the assumption that opening up innovation was principally about maximizing the returns on R&D by pursuing more than one path to market. This turned out to be the easy bit. It wasn't very hard, after all, to convince the CEO that licensing out under-utilized technologies was worthwhile, as long as the program could bring in revenue, or perhaps turn R&D into a profit center.

As companies climb up the open innovation learning curve, however, they soon discover that the real value of an open market for innovation lies in getting access to external ideas that can fill performance gaps or fuel their product pipelines. Phil Stern explains that yet2.com's offering has evolved accordingly. It's no longer just a matter of, "How do I off-load underutilized IP," says Stern. Increasingly, it's, "How do I access technology from the outside to feed this very hungry growth engine."

Ask anyone at P&G why the company is leading the consumer products industry, and they will tell you it's because the company is constantly innovating. But with the pace of innovation in the industry having doubled in five years, an army of nine thousand researchers is no longer sufficient to keep P&G at the forefront of innovation in consumer products. Though nine thousand may sound like a lot, it's actually a tiny share of the researchers in the field. In fact, for every P&G researcher there are two hundred scientists or engineers elsewhere in the world who are just as good. That's a total of 1.8 million people whose talents it could potentially tap into.

When P&G launched its "connect and develop" initiative to help tap this vast reservoir of talent, the idea wasn't to replace these nine thousand researchers, but to better leverage them to drive growth and innovation. As part of the initiative, business-unit leaders have been asked to source 50 percent of their new product and service ideas from outside the company by 2010. It's a fairly radical step, but A. G. Lafley judges that it is so critical to P&G's future that he has made "proudly found elsewhere" a mantra for the company.

Constant change and growth are now the essence of survival. "Most mature companies," says Larry Huston, "have to create organic growth of five to seven percent year in, year out." Relying on internal capabilities to produce that growth rate may have worked when P&G was a $25 billion company, he argues. But today it's worth $70 billion. Organic growth of 6 percent, for example, is the equivalent of building a profitable new $4 billion business every year!

Other mature, innovation-based companies face the same dilemma. The potential for growth is out there—but it's distributed across thousands, perhaps millions, of individuals, organizations, and firms. Small and medium-size businesses, universities, and even individuals are increasingly important sources of innovation. They are hungry to join up with loosely coupled business webs. But getting access to all of these ideas and organizations is tough. It's a bit like looking for a list of local plumbers without the Internet, or even a phone book. That's where ideagoras come in—they help link all of these individuals, companies, and organizations together by establishing connections and facilitating transactions between buyers and sellers of ideas and technology.

Companies that think they can fall back on outsourcing and off-shoring are missing the bigger picture. Contracting out R&D work to a handful of lower-cost providers might help companies reduce costs and increase the number of researchers at their disposal. But it won't unleash a wellspring of collaboration and innovation the way millions of seekers and problem solvers could in a global ideagora.

Companies still need internal research capabilities to ensure they can add value to external ideas. But, more important, they need the equivalent of Goldcorp's open approach to exploration or IBM's foray into open source—they need highly permeable corporate boundaries. By exposing problems to the world they can engage and cocreate with the most uniquely qualified minds to solve them.

Through connect and develop, for example, P&G collaborates with organizations and individuals around the world. It scours the globe for proven products and technologies that it can improve, scale up, and market, either on its own or with its business web. When the company finds those good ideas, Huston says, it brings them inside, where it can capitalize on internal capabilities to enhance or append the offering. Successful products such as Olay Regenerist, Swiffer Dusters, and the Crest Spin-Brush are just 3 of hundreds of products that P&G has brought to market through connect and develop in the past few years.

Not only can ideagoras enable companies like P&G to innovate well beyond what they could muster internally, they help them hone their true value-adding capabilities and avoid reinventing the wheel. So, for example, when P&G set out to launch a new line of Pringles potato chips with trivia questions and animal pictures printed on each chip, it quickly discovered that producing sharp images on thousands upon thousands of chips each minute was a highly complex endeavor.

In the past, P&G would have dedicated considerable internal resources to figuring this out, and perhaps even partnered with a printer company that could help devise a workable process. But with an ideagora, P&G could do better. It formulated a paper describing the technology and tapped its global network to see if there was a uniquely qualified mind that could solve the problem. A solution popped up in a small bakery in Bologna, Italy, where a university professor was printing edible images on cakes and cookies. He'd cooked up an ink-jet method in his bakery, and it looked like it would solve P&G's problem. So P&G acquired the technology and quickly adapted it to its requirements.

Huston says P&G was able to launch Pringles Prints in less than a year, and for much less than what it would have otherwise cost. Applying this same philosophy in all its business lines enables P&G to focus on areas where they're going to be world class and to source technology in areas where they don't need to be leading edge.

The strategy has paid real dividends. In a subsequent *Harvard Business Review* article Larry Huston and Nabil Sakkab reported how opening up innovation had truly transformed the company. "Today," they wrote,

> *more than 35% of our new products in market have elements that originated from outside P&G, up from about 15% in 2000. And 45% of the initiatives in our product development portfolio have key elements that were discovered externally. Through connect and develop—along with improvements in other aspects of innovation related to product cost, design, and marketing—our R&D productivity has increased by nearly 60%. Our innovation success rate has more than doubled, while the cost of innovation has fallen. R&D investment as a percentage of sales is down from 4.8% in 2000 to 3.4% today. And, in the last two years, we've launched more than 100 new products for which some aspect of execution came from outside the company. Five years after the company's stock collapsed in 2000, we have doubled our share price and have a portfolio of 22 billion-dollar brands.*[5]

If that's not serious payoff from the new engage and cocreate model, then we can't say what is.

HARNESSING IDEAGORAS

The closed and vertically integrated approach to innovation that companies employed for generations came with a requisite set of capabilities. Successful firms had everything from the in-house brainpower to the management and marketing muscle to compete in national and international markets. Above all, they had to be good at managing a linear innovation pipeline that chewed on basic scientific research and spat out marketable products and services.

With the emergence of global ideagoras, companies can pursue a wider set of strategic possibilities. Firms can choose to acquire outside ideas and technologies instead of developing them in-house, or they can choose to license out their technology instead of (or in addition to) commercializing the goods.

To make the most of this broader strategy, companies will need to nurture new collaborative capabilities. "The simple truth," says A. G. Lafley, "is that the most adaptive, agile, and responsive companies are almost always the most in touch. The companies that are the most in touch tend to be the most collaborative. And the most collaborative—the companies that are the best at creating, finding, and reapplying great ideas—are those that sustain growth over the long term."

Learning how to create, find, and reapply great ideas in a global ideagora means turning the R&D organization on its head. Some of the big changes include refining its approach to intellectual property, sharpening its external radar, and creating an R&D culture that supports the acquisition of external ideas and technologies. Intermediaries like yet2.com and InnoCentive, meanwhile, will need to work alongside their clients to help improve the market's liquidity and build business norms and practices that foster collaborative innovation. In the remainder of this chapter, we explore some of these key issues.

Creating Liquidity

A little over twenty years ago, the idea of appointing a chief executive for information technology was still novel, even laughable. Information technology was like plumbing—so why on earth would you appoint the equivalent of a plumber to the level of chief executive?

That was before companies realized that IT was becoming a source of competitive advantage. When it became clear that IT could be used to build more efficient and effective business models and structures, executive-level IT leadership became a necessity.[6]

Today we are at an equivalent stage in the path toward collaborative innovation. Most firms develop 90 percent or more of their differentiating technology in-house. Indeed, with 35 percent of its innovations coming from outside its walls, P&G is by far the exception. Others, like IBM and Lilly, are leaders in their respective industries, in that they have gone farther than most in experimenting with new ways to source innovations from outside their corporate boundaries.

The upside is that there is tremendous room for innovation marketplaces to grow. Companies that move early to hone their skills in these marketplaces will enjoy considerable competitive payoffs for some time to come.

The downside is that liquidity in nascent ideagoras is typically poor. In other words, there are too few buyers and sellers, and therefore too few transactions to make the marketplace a vibrant source of great connections and innovations. The underdeveloped state of the marketplaces inhibits the ability of companies to find what they are looking for externally, which, in turn, puts a damper on open innovation. It's a classic chicken-and-egg problem.

Expectations were set very high when marketplaces such as yet2.com and InnoCentive first came along. "We were initially of the belief—as were many business-to-business marketplaces—that we were going to hit a tipping point very early on," says yet2.com CEO Phil Stern. Few realized just how tricky building marketplaces would turn out to be. It's easy to get carried away by the successes of eBay. Exchanging technologies is a much more complex ballgame.

Consider the problem from the perspective of a typical licensing executive or R&D professional. "A technology buyer at a typical large company," says Stern, "first needs to describe the technology need or benefit that will enable the intended product performance. He or she then needs to search for technologies that provide that benefit and evaluate the incoming candidates. Only then can he or she attempt to get approval to purchase or license it. That piece turned out to be a lot more difficult than we thought."

Poor liquidity makes a hard job even harder. "It's pretty hard to say to your boss, 'Well, I found this one solution to our problem,'" says Stern. "His or her boss might come back and say, 'Well, how do you know it's the best one? How many others are out there? Are there solutions that may be farther along than this one?'" Without a liquid market presenting lots of competing options it is very hard to know.

Unlike eBay, where goods are exchanged for cash, transactions in technology markets tend to be more complicated. Many transactions, for example, include a significant incentive for companies that are transferring the technology to help the acquiring company be successful with it. Transactions between large firms and start-ups are often structured such that the large firm transfers the technology in exchange for equity in the new company. Other deals grant exclusivity to operate in a certain geography or for a certain time period. In 99 percent of the cases, getting to a deal requires a fair amount of creativity.

The bottom line is that there is a lot more to transferring ideas and technologies than exchanging legal documents. When licensing out, companies should expect potential partners to require technical information, demonstrations, samples, and testing to verify that the technology can do what they need it to do. They may even need to provide technical assistance and other services as a prelude to the actual licensing.

When bringing in external ideas, companies can never assume that ready-to-go ideas are truly ready to go. Unless one is talking about generic scientific knowledge or technologies that are easily understood or codified, there can be substantial costs involved in operationalizing external knowledge in its new context. In such cases, technology transfer can be as expensive and time consuming as independent R&D.

Apart from providing a common platform for transactions, intermediaries like yet2.com and InnoCentive play an important role brokering deals and helping increase liquidity. For example, yet2.com helps its clients identify an array of technological solutions so that they feel confident that a full range of opportunities have been explored. It also helps clients succeed in closing the deals. "Standing on the sidelines and watching our clients go to hand-to-hand combat is dangerous," says Stern, "so we put more and more emphasis every day on facilitating those transactions." Its efforts in brokering deals have seen the deal closure rate double for two years consecutively.

InnoCentive plays a similar value-adding role. In brokering communications between the seeker and the solver, for example, InnoCentive makes sure to shield the identity of the client and the nature of the ultimate application. Its managers also ensure that the solver's work is authentic and that a seeker will not unwittingly use technology or solutions to which they do not own the intellectual property rights.

Though the market for innovation is clearly headed toward more liquidity, no one can point to when. "We will be a big beneficiary, and hopefully a big cause of that liquidity, as time goes on," says Stern, "but every individual client has to prove these economics to itself."

The ultimate driver of liquidity will be the successes won by trailblazers like Procter & Gamble. If more P&Gs start racking up billion-dollar wins from ideas they have sourced externally, the economics will be self-evident. "P&G and others will be using this capability to drive innovation in a way that others just can't compete with," says Stern. "Other companies will simply have no choice but to figure out how to be successful at it."

Instilling the Culture

Companies still need to break down deep-rooted biases that inhibit them from seizing opportunities to open up innovation. Many firms are just now coming to the realization that they can turn some of their underutilized assets into new and lucrative revenue streams.

Like many companies, General Electric used to see IP as a defensive tool. "We didn't systematically enforce it or try to generate revenue from it," says Dave Christensen, a licensing executive. "We just kept it in reserve for protection."[7]

After years of watching companies like AT&T, IBM, and Texas Instruments build up healthy licensing businesses, mind-sets are beginning to change. In GE's case, Jack Welch simply had to take note of how much money IBM was making from its licensing business before a mandate for change was handed down. "The legal group came to believe that IP could be a profit center and pay all of our patent maintenance, prosecution, and preparation fees . . . [while] the technology group realized that revenue from IP could be used to fund future technology growth and new projects," says Christensen.

Even so, many companies seeking to generate revenue from their research lack an effective organization for directing these activities. Those that do typically anchor activities like technology licensing in the legal department. Consequently, licensing activity tends to be reactive, not proactive.

If the sell side of the global market for ideas and technology seems a bit rusty, the buy side is like tetanus. People have been talking about the "not invented here" syndrome for years, if not decades. And for the most part the culture still hasn't changed much. The modus operandi of R&D departments is to invent, not to acquire outside ideas. "Ask a scientist to do problem solving," says Alf Bingham, "they'd say, 'You bet, that's me, a problem solver.'"

Add a hint of arrogance to the equation, and you can easily see why many companies haven't moved quickly to hunt for innovation outside the firm. "We used to say that we could invent our way to anything," says Robert Hirsch of DuPont, whose comments echo those of many seasoned R&D professionals. Things are changing at DuPont these days. The senior leadership realizes that DuPont can't and shouldn't do everything itself. "It's a question of cost and speed," he says. "You may be able to acquire something externally that provides know-how at a much lower cost."[8]

Still, these kinds of cultural changes are a bit like painting over 1960s wallpaper instead of knocking down walls and rebuilding. As Alf Bingham explains, "It requires a lot of trust to believe that you can accomplish your goals by relying on freelance scientists to come up with solutions. Most people at big companies are not ready to entertain that idea." Indeed, our research suggests that managers are comfortable with a change in the players (i.e., outsourcing), especially if it's economically advantageous. But they're not as comfortable with a change in the R&D business model.

Companies who want to change the game with open innovation should be driving for a qualitative change in approach, not just incremental tinkering. R&D organizations in large companies especially should be striving to add value by knowing what the important questions are and helping to integrate the final solutions. Let most of the research activity reside in scientific networks. And let firms solve the tough business and resource orchestration problems.

Organizing loosely coupled innovation webs to capture the serendipity and diversity of innovations and scientific progress is where the real value is. Flexible and distributed human capital networks like InnoCentive will make it easier to execute projects with ad hoc teams that converge temporarily to assemble the intellectual, financial, and physical assets required to bring products and services to market. When that happens we may scarcely recognize the corporate form that emerges to facilitate this kind of ephemeral capitalism.

In the short term, staying power, CEO-level leadership, and a commitment to appropriate staffing, incentives, and organization will be critical success factors. "Open innovation efforts can die like a flash in the pan," says Stern, "because some senior executive says, 'I'm giving you two people and an external budget of ten thousand dollars, and we'd like to see ten deals in the first year.' It just doesn't happen. It takes quite a bit of time for these initiatives to mature."

P&G's Larry Huston reminds us that no amount of idea hunting on the outside will pay off if, internally, the organization isn't behind the program. "Once an external idea gets into the development pipeline it still needs R&D, manufacturing, market research, marketing, and other functions pulling for it," he says. Moreover, there needs to be very senior level support, ideally from the CEO.

Creating appropriate incentives is equally part of the recipe. Many firms reward R&D professionals for generating patents. Employees should also be rewarded for spotting and acquiring external ideas. Basing rewards on the ability to get successful products to market quickly makes the incentive system neutral to whether ideas originate inside or outside the firm.

When it comes to profiting from underutilized assets many companies are tempted to pour the revenues into general corporate coffers. The incentives to participate, however, are much stronger if revenues flow to the bottom line of the groups that invented it. When business-unit leaders see licensing income starting to make their P&L statements look rosy, they'll be knocking down the door to get involved.

Harvesting External Ideas

The ability to harvest external ideas starts with a keen sense of what you are looking for. The Internet may have shrunk distance and time, but the world of ideas and technologies is still a vast open space. Firms will need clear goals and guidelines to avoid getting lost or sidetracked by unplanned journeys through a maze of opportunities.

All journeys to the technology hinterland should begin with some basic stock taking. What do our customers need today? What will they need in the future? How can we complement or add value to our existing products and services? What new market opportunities present the greatest opportunities for growth? As we develop new ideas, what can we deliver internally? What should we source externally? Are there exciting new clusters of innovation happening that we can tap into? Where can we work closely with partners to create even more value? Which external acquisitions can we turn into deeper and broader collaborations?

Large companies with diversified business lines need to think through these questions carefully in order to identify the best external candidates. P&G, for example, now spends close to $2 billion on R&D across 150 science areas and 300 brands that span multiple product categories. To fine-tune its search the company directs its external surveillance activities to three environments. It begins with a list of its top ten customer needs that include broadly defined goals that are subsequently boiled down to solvable scientific problems. Next it creates a list of adjacencies—new products or concepts that can help it take advantage of existing brand equity. Finally, it utilizes "technology game boards," a sophisticated planning tool that allows P&G planners to assess which key technologies might be central to several overlapping product categories or brands, and thus make good candidates for strengthening.

Larry Huston compares the whole exercise to a "multileveled game of chess." Yet, even with this highly refined approach to filtering opportunities only one in one hundred external ideas identified by P&G ends up in the market. In fact, Huston says, "To get over one hundred products in the marketplace we had over two million briefs sent out!"

The natural complement to a keen sense of purpose is an astute external sensing capability to identify new markets, new technologies, and emerging competitive threats. Though no one can know everything about a given opportunity or threat, scenario planning and collective intelligence gathering can aid external sensing and sense making. Plugging into Internet-enabled marketplaces such as yet2.com and InnoCentive will ease the process of discovering and transacting for solutions.

Despite the necessity of exploring electronic markets, limits to liquidity make it unwise to rely on marketplaces alone for access to external ideas and technologies. P&G, for example, uses ideagoras, but it also employs a network of technology scouts that literally scour the globe for new consumer products and technologies. According to Larry Huston, these are senior people that develop P&G's list of technology needs, identify external opportunities and connections, and actively promote these connections to decision makers in P&G's business units. Their methods blend aggressive mining of the scientific literature, patent databases, and other data sources with physical prospecting for ideas, including surveying the shelves of a store in Bangkok or combing product and technology fairs in Beijing. Indeed, a technology entrepreneur, exploring a local market in Osaka, Japan, discovered what ultimately became the Mr. Clean Magic Eraser, a stain-removing sponge that has achieved double its projected revenues.

Universities, partners, suppliers, customers, and communities of practice all represent additional sources of input. Technology companies like HP, Intel, and Google, for example, leverage worldwide networks of university research labs that are advancing the state of the art in areas of strategic importance. In addition to universities, these companies tap open source communities as a development partner and a source of new ideas and technologies. A growing number of companies, meanwhile, are engaging their customers in product design, and listen closely to what their target demographics are saying in their weblogs and podcasts.

Sensing opportunities lead to the need for creative design—conceiving the ultimate customer offerings and the technical/business architectures for delivering them. This is where the really hard work of translating what appears to be exciting external ideas into the internal company gearing, pipeline, and culture. This means thoroughly evaluating a technology's potential, selling it into the relevant business units, helping the business unit realign their product road maps and infrastructure if necessary, and conducting research into the product's business potential.

Creative design also includes the ability to execute smart decisions about acquiring IP (what/how) in the first place and partnering (who/when/how) to design, assemble, and deliver the final value. In any cross-licensing scenario (i.e., where rights to IP are swapped to avoid transfers of cash or royalties) firms need to carefully assess what they are putting on the table in order to access the IP they need to fulfill their grand designs.

In the creative design process, firms should adhere to the principle that markets allow for an increasingly specialized division of labor. Use ideagoras to play to your core strengths. Concentrate R&D in those areas where you have the greatest competitive advantage in developing valuable innovations, and use ideagoras to acquire the rest.

While reaping the benefits of specialization, however, it is also important to remember that absorbing external technology or IP depends on the ability to relate what you learn to what you already know. Internal R&D and external acquisitions are complements, not substitutes.

Finally, with all of the promise that ideagoras offer, there is one further word of warning. Companies that engage actively in ideagoras must remember that markets are great levelers. Whenever a market is open to all qualified comers, competitive advantage will not flow directly from participation in that market. Competitors can simply enter the same market. This fact suggests that the domains in which unique value can be built are likely to shrink as ideagoras become increasingly open and competitive.

As markets for know-how and intellectual property grow, competitive advantage will depend on the combined ability to create, transfer, assemble, integrate, and exploit knowledge assets. Superior technology alone will not produce competitive advantage. With just a little time and effort, most technology can be invented around.

If there is a saving grace, it will be the speed and skill with which companies can harness marketplaces to develop new customer offerings. Organizational agility allows firms to move strongly and rapidly into a latent market when new IP emerges in the marketplace. More than ever, this means leveraging both internal and external competencies through partnerships and alliances, many of which can be solidified by acquiring and trading IP. Firms that have these dynamic capabilities are most likely to be entrepreneurial, with flat hierarchies, clear vision, effective incentives, and employee autonomy.

Getting the Right Ratio

Reaching an optimal ratio of internal to external innovation presents a unique challenge to companies as they strive to tap the power of ideagoras. How much external technology is too much? How little internal R&D is too little? P&G has publicly declared a target to obtain 50 percent of its new innovation from outside the company before 2010. Is that enough or too much? Should all companies across all industries strive to reach similar milestones? Navi Radjou, Forrester's leading innovation expert, has suggested that companies shift their innovation mind-set from "everything invented here" to "nothing invented here." While we are optimistic about the potential for open innovation, we think this goes too far, perhaps even dangerously so.

Companies that invent get an opportunity to shape the future. They don't need to invent everything in-house, own all of the IP, or employ all of the people that contribute to their innovation webs. But they do need to add some value to the ecosystem. And they need to be able to add real value to any IP they acquire on the open market. Companies who lose their edge, or fail to compete in product differentiation, narrow the parameters on which they can compete to cost and branding alone.

Not all companies need a war chest of IP, but those that have none weaken their hand in the very negotiations required to get access to external IP in the first place. Without ideas and inventions of their own they may have little to use as bargaining chips in licensing and cross-licensing negotiations. They are forced by default to give up cash and/or royalties instead.

Moreover, there are tasks that external IP marketplaces are poorly equipped to handle. InnoCentive, for example, works well for bounded problems—i.e., problems that have easily definable parameters and outcomes. But companies are always going to need a core group of R&D people to ask the right questions, draw up strategies, source the external inputs, and help commercialize the end products. There will also be problems that need to be worked on collaboratively, where there is a little bit of fumbling around that's required under the direction of the people who understand the issues.

To be fair, Radjou's broader point is that you don't need to invent to innovate. He points to the example of Dell, a company that invests little in R&D compared to companies like HP and IBM. Yet Michael Dell has been trying to overtake HP in the printer business, despite the fact that HP invests $1 billion of R&D in its printers alone. Dell's plan: License printer technology from HP's rival Lexmark and transform it into Dell-branded printers using its world-class supply chain.

Few doubt Dell's ability to harness its supply chain processes to shed costs. Perhaps Dell's lack of in-house printer technology is adequately compensated for by its competencies in supply chain management and distribution. But the strategy raises some important questions.

Lexmark's technology is already a poor second-best to HP, so is it really worth licensing? And how will Dell keep up with the pace of improvement in printing technology without investing anything in R&D? Indeed, does Dell need to keep up in R&D if it's ultimately satisfied pursuing the lower tier of the market?

Some answer that Dell bundles its newly branded printers with its computers at a loss just to get a piece of the lucrative replacement ink cartridge market. Others claim that Dell's plan has more to do with draining HP's profits (about 60 percent of which come from printers and ink), profits that HP has used to bankroll price wars with Dell in PCs and servers (where Dell is still number one). Either way, Dell's strategy is all about beating HP on price. With that approach, it can't afford to do much about advancing the state of the art of printer technology.

Meanwhile, over in San Jose, HP is betting it can fend off any move toward commoditization, particularly in the lucrative enterprise market. The company has been working on digital printing technology it claims will reduce corporate printing costs by as much as 30 percent, a figure that Dell will not beat without reinvesting profits in R&D and creating its own proprietary technology. With HP taking aggressive steps to trim its R&D costs and still miles ahead in market share terms, it seems unlikely that this is a battle that Dell will win.

The lesson is that just as companies carve out competitive advantages in different ways, they can emphasize different ratios of internal to external research and still succeed in the marketplace. Not every company is going to be good at research. Not every company needs to be. Like Dell, they can acquire external technologies to drive their innovations. But then they will need to excel in other areas, like cutting costs or orchestrating a value chain. They'll also need to choose their offerings wisely, and shouldn't expect to dominate markets in which the technology is still rapidly improving. In this particular scenario, licensing just isn't likely to give you that cutting edge.

The Portfolio Approach

Marketplaces like yet2.com and InnoCentive are a growing force in the new landscape. By no means, however, do they offer the only path to open innovation. Smart firms will harness a portfolio of approaches ranging from corporate ventures to customer cocreation to peer producing value in open communities to developing innovations within proprietary networks of partners and suppliers.

Take IBM, a company with decades of experience in harnessing innovation markets. With over a billion dollars in annual licensing revenue, IBM is regarded by many as a leader in generating cash flow from its war chest of patents. Even then, Joel Cawley, a senior strategist at IBM, admits that IBM still has a very large library of assets that are undercontributing to its business performance. "One of our central focuses now," says Cawley, "is to figure out how to more effectively exercise those assets to drive our business."

The important point for Cawley and other IBM strategists is that technology marketplaces are just one tool in the kit bag. There is a much broader portfolio of initiatives that will put IBM's ideas and technologies into practical use and generate value for the company. Placing IP in a protected commons where open source communities can improve and build on it is another method that is gaining credence across the company. As described in Chapter 3, collaborating with the open source community has already generated big wins with Linux and Apache, and today open source is a keystone in the company's technology strategy. IBM also partners with venture capital firms to put its portfolio of technology and know-how into the hands of start-up companies that may one day evolve into promising businesses.

Like IBM, P&G also uses a portfolio of open innovation approaches. Ideagoras such as yet2.com and InnoCentive are an important part of the equation. The company has even invested in or helped create other marketplaces like NineSigma and YourEncore.

Equally important, however, are P&G's suppliers. The company's top fifteen suppliers have a combined R&D staff of fifty thousand and represent a significant source of innovation. "We have people in the labs of our suppliers, we have their people in our labs here," says Larry Huston, "and sometimes you can't tell the difference." If P&G wants to develop detergents that will work well with energy-efficient washing machines that use less hot water, it turns to its partners and suppliers who may well have the answer. P&G is also working on initiatives to encourage more supplier-to-supplier collaboration so that technology briefs will be shared across the ecosystem in a bid to speed up innovation.

Working in close collaboration with suppliers to produce new products and services is becoming a mainstay in most industries. Unlike the old days, where lead companies handed down detailed specs and expected suppliers to conform, the new modus operandi is based on collaboration and cocreation from the design stage through to manufacturing. Like P&G, many companies now swap personnel and work side by side in the lab.

The key point is that there is a diversity of approaches to opening up innovation both in terms of the paths companies take to get ideas and technologies to market and the sources from which they draw in external inputs. Successful companies will harness a wide selection of these strategies, if not all of them. Though we have focused this chapter on the potential to harness marketplaces, you will find much more on peer production, supplier collaboration, and customer cocreation in other chapters.

Pushing the Envelope

The last and best piece of advice is to push the envelope. Ideagoras lower the costs of communicating, collaborating, and transacting and could very well revolutionize the way firms conduct R&D. Companies that learn how to harness ideagoras will divest themselves of noncore activities and conserve their resources for the cutting-edge challenges and opportunities. Smart firms will build their R&D organizations around a core of question askers and outsource most of the problem solving. But rather than put down bricks and mortar in China, they can just tap a marketplace like yet2.com or InnoCentive.

In addition, there is a great deal of room for these nascent ideagoras to improve the way they deliver their services. InnoCentive, for example, could spur some pretty profound changes if it looked and behaved a little bit more like the open source communities behind Linux and Apache. InnoCentive researchers, for example, do not naturally coalesce into large groups focused on collaborating to solve a single problem. Nor does InnoCentive offer the openness and transparency of open source software. Seeker firms can cloak their identities, and solvers may never get personal credit for their contributions. InnoCentive could increase contributor activity and loyalty by addressing these challenges.

In some cases, collaborative problem solving is evolving spontaneously. Graduate students at Duke University are forming InnoCentive Solvers Clubs, and solvers in German and Chinese universities collaborate on a variety of interesting problems. InnoCentive has already signed agreements with leading universities in India and China, where student teams are uniting to gain valuable experience, recognition, and income.

By expanding the tools available to users to manage rights and communicate with other site users, InnoCentive could begin to ignite more vital, self-organizing behavior among seekers and solvers. Teams of solvers, for example, could coalesce into firms or self-organize into ad hoc freelance organizations. Indeed, where problems are highly integrated, it may be preferable to offer problems to skilled external teams rather than or in addition to posting them in an open market. The fact that InnoCentive structures its problems in a modular way already means that there are opportunities for individuals and organizations to build a business around this model of innovation.

ON A JOURNEY TO THE NEW INNOVATION

Twentieth-century firms brought innovation within their boundaries for good reasons. R&D was closely aligned with the firm's existing proprietary product and process technologies, its strategy for staying ahead, and its market opportunities as it saw them. Effective lab work often required not just industry-specific knowledge, but also firm-specific knowledge. Much of this knowledge was uncodified, and a great deal of learning and refinement occurred in the course of doing business, not in the lab.

Moreover, the difficulty of knowing in advance how an R&D project will turn out made it necessary to rethink and respecify objectives frequently. All of this complexity made it inherently difficult to pursue innovation through arm's-length relationships and contracts.

In-house R&D also made it much easier to control intellectual property. The motivation for the R&D and the content of the project itself was rightly regarded as proprietary. Since firms profit from R&D largely by exploiting a head start, the details of R&D needed to be kept private until ready for practice.

Internal R&D will still be important in the new world of ideagoras. As always, tomorrow's technology will still largely grow out of today's. As a consequence, durable competitive advantages in R&D-intensive industries will still be rooted in the growth of deep domain-specific knowledge. Firms will still need to invest resources in internal R&D to be able to recognize a commercial opportunity and exploit it quickly. Companies that expect to be players will still require broad, high-quality portfolios of proprietary IP for trading and cross licensing.

But in-house innovation alone will not be enough to survive in a fast-changing and intensely competitive economy. The days when companies divided the national pie in a wading pool of well-mannered rivalry are long gone. Today's global firms sink or swim in an ocean roiling with cutthroat competition. And plummeting collaboration costs enable them to reach outside their boundaries for talent.

Throwing more money into internal innovation will deliver fewer and fewer breakthroughs as the costs of R&D escalate. Companies that don't source a growing proportion of new product and service ideas from outside their walls will find themselves unable to sustain the level of growth, agility, responsiveness, global savvy, or creativity they require to compete in today's environment.

Increasingly, the corporate R&D process must look two ways: toward its internal projects and competencies, and toward the external marketplace to leverage new IP and capabilities. Innovation must extend beyond the boundaries of the firm to the very fringes of the Web, where companies will engage with customers and a dynamic network of external collaborators. Ideagoras are the place where companies can tap a wealth of new ideas, innovations, and uniquely qualified minds. Think back to the Goldcorp Challenge, and use ideagoras to find your next eight million ounces of gold.

5. THE PROSUMERS
Hack This Product Please!

O ur community is an opportunity to take a look at the rules that govern society, and to the extent that we are able, rewrite them as best seems to fit us," said Philip Linden to author and Stanford law professor Lawrence Lessig, as they sat down for a Q&A session in the amphitheater of a fascinating new settlement. Lessig nodded his head in agreement. A bit of a folk hero in these parts, Lessig was making a specially scheduled appearance to discuss his books, *Free Culture* and *The Future of Ideas*, with an ensemble of several hundred of its residents. Philip Linden, his host, was among the original homesteaders in this pioneer community.

"For those here who don't know, Lawrence has affected the history of our community already," said Linden, as he introduced Lessig to his compatriots. "We had a meeting in 2003 to think about our future, and Lawrence was kind enough to attend, and to give us his thoughts on IP, land, and how things should be. Shortly thereafter we gave IP rights to creators and switched to our system of land ownership."

"Bravo," replied Lessig, taking in the scenery.

"As every free society has discovered," said Linden, "we have realized, more and more over time, how much our community is a developing nation, and how, if we want to succeed, we must make the choices that advance us all."

"That's why the people here are so important to this debate," Lessig said encouragingly. "You have got to make the clueless politicians aware of what nineteenth-century law is doing to the twenty-first century," Lessig said, getting more animated. "They don't get it. They think they're stopping 'pirates' when they stop all sorts of creativity."

As interesting as Lessig's comments were, it was the venue in which he made them that is truly remarkable. Despite appearances, Lessig and his host are not members of a cultlike hippy enclave in a remote part of New Mexico. Lessig was appearing, not in person, but as an avatar in a virtual stadium, and the hundred-plus residents who came to listen in were all virtual avatars as well. All of them were participating in a virtual world of their own making—a massively multiplayer online game (MMOG, for short) called Second Life, where millions of participants socialize, entertain, and transact in a virtual environment fabricated almost entirely by its users.

In fact, Second Life residents are far more than just "users." They take on virtual identities, act out fictitious roles and activities, and even create virtual businesses that earn 200 residents upwords of $5,000 each. *Business-Week* writer Robert Hof aptly calls Second Life, "[T]he unholy offspring of the movie *The Matrix*, the social networking site MySpace.com, and the online marketplace eBay."[1]

One player, who goes by the pseudonym Anshe Chung, runs a virtual real estate development company and residents pay Linden dollars, the in-game currency, to buy or rent the ornate virtual homesteads her company designs. Even at three hundred Linden dollars to one buck, Chung does some brisk business. Chung's holdings of Linden currency and virtual real estate now surpass the equivalent of a quarter of a million dollars. She says, "This virtual role-playing economy is so strong that it now has to import skill and services from the real-world economy."[2]

Players like Anshe Chung, and indeed all players in Second Life, are not just consumers of game content; they are at once developers, community members, and entrepreneurs—and, like Chung, a growing number even make their living there. This means Second Life is no typical "product," and it's not even a typical video game. It's created almost entirely by its customers—you could say the "consumers" are also the producers, or the "prosumers." After all, they participate in the design, creation, and production of the product, while Linden Labs is content to manage the community and make sure the infrastructure is running.

In his 1996 book, *The Digital Economy*, Don introduced the term "prosumption" to describe how the gap between producers and consumers is blurring.[3] Though many now recognize the significance of this development, most still confuse prosumption with "customer centricity," where companies decide what the basics are and customers get to modify certain elements, like customizing your vehicle on the showroom floor. Even TiVo, which makes you "the programmer" (i.e., the person who sets the TV schedule), is not as exciting as producing your own homegrown content. In our view, all this customer centricity is pretty much business as usual.[4]

This chapter describes a new model of prosumption, where customers participate in the creation of products in an active and ongoing way. As in Second Life, the consumer actually co-innovates and coproduces the products they consume. In other words, customers do more than customize or personalize their wares; they can self-organize to create their own. The most advanced users, in fact, no longer wait for an invitation to turn a product into a platform for their own innovations. They just form their own prosumer communities online, where they share product-related information, collaborate on customized projects, engage in commerce, and swap tips, tools, and product hacks.

By learning how to harness a prosumer community for competitive advantage, Second Life originator Linden Labs has broken most of the conventional rules for building a multiplayer video game and set the standard for customer innovation in all industries. It's not yet the largest MMOG, but it is growing fast. As of January 2010, Second Life had about 18 million registered users.

While most multiplayer games are themed and scripted by a handful of internal designers, Linden Labs has gone to the other extreme, opening up its gaming environment in radical new ways. Second Life has no preset script, and there are few limitations on what players can do. Residents create just about everything, from virtual storefronts and nightclubs to clothing, vehicles, and other items for use in the game. In fact, Linden Labs produces less than 1 percent of its content and now gets up to 23,000 hours of "free" development effort from its users every day.

Users don't give up all of this labor for nothing. In Second Life anything a resident creates is theirs. While some multiplayer games forbid real-world trades of virtual goods, the practice is sanctioned, even encouraged within Second Life. Industry powerhouses like Sony Online Entertainment president John Smedley say giving users IP rights would be like "getting a gym membership and saying you own the equipment." But for Linden Labs it's all about building a giant, freewheeling, customer-driven economy that currently turns over an estimated $100 million per year.

Second Life's prosumptive approach to building a business offers advantages that tightly controlled business models can't replicate. It makes big impacts with fewer resources. It scales in ways that centrally designed systems cannot. It benefits from positive feedback loops that are difficult for competitors to reverse. It innovates more rapidly, and engages stakeholders in loyal communities, because the players create the rules of the game, own their IP, and even volunteer to provide customer support.

Companies should follow Linden Labs' lead in building a "product" that invites and enables customers to collaborate and add value on a massive scale. These opportunities to add value should extend throughout the product life cycle, starting with design and extending to aftermarket opportunities for customer-driven commerce and innovation. This chapter will explain the process with a series of cases that explore how self-organizing prosumer communities introduce both lucrative opportunities and grave new threats to companies.

For the managers who are wondering whether this is serious, Second Life also sends a warning. In the same way that Second Life is an infinite platform for customer innovation, not a product, this new generation of prosumers treats the world as a place for creation, not consumption. This new way of learning and interacting means they will treat the world as a stage for their own innovations. Just as you can twist and scramble a Rubik's Cube, prosumers will reconfigure products for their own ends. Static, immovable, noneditable items will be anathema, ripe for the dustbins of twentieth-century history.

CUSTOMERS AS CO-INNOVATORS

The idea that the people who use products should have input into their design and production is not entirely new. There have been many episodes of user-driven creativity in the history of invention as scholars such as MIT professor Eric von Hippel have pointed out. In early nineteenth-century England, Cornish steam-engine makers collaborated openly with mine owners to improve the efficiency of the steam engines used to pump water out of the coal mines. In the United States, the mass production of steel in the 1870s and the invention of the personal computer in the 1970s were both preceded by long periods of open tinkering within the community of users and technicians. In those cases, technology was pushed into applications, and new industries emerged rapidly, because technical people openly discussed and shared what they were working on.

Other research has shown the great importance that hobbyists and "amateur" creators play in advancing technology. A casual look through the pages of a 1950s edition of *Popular Science* reveals a vast treasure trove of amateur innovation in fields ranging from electronics to scientific instruments and mechanics. Even the Model T (the car you could get in any color you wanted as long as it was black) was subject to intense customization by its customers—a trend that continues today in increasingly large communities of auto enthusiasts and aftermarket specialist shops, and of course, on MTV television shows like *Pimp My Ride*.

Despite this rich history of customer innovation, most companies consider the innovation and amateur creativity that takes place in communities of users and hobbyists a fringe phenomenon of little concern or value to their core markets. Firms often resist or ignore customer innovations. It took car manufacturers more than a decade to "invent" the pickup truck, after American farmers had spent years ripping the backseats out of their vehicles to make room for their goods and tools. Even when customer innovations look promising, most companies' internal processes have been too rigidly adapted to the manufacturer-centric paradigm to make use of them.

This reticence is set to change, however, as two forces converge to upset the status quo. One, as we have already explained, is that customers use the Web as a stage to create prosumer communities, so what was once fringe activity is increasingly out in the open. Second, companies are discovering that "lead users"—people who stretch the limits of existing technology and often create their own product prototypes in the process—often develop modifications and extensions to products that will eventually appeal to mainstream markets.[5] In other words, lead users serve as a beacon for where the mainstream market is headed. Companies that learn how to tap the insights of lead users can gain competitive advantage.

BMW, for example, employs thousands of R&D professionals and has an entire shop in Silicon Valley dedicated to producing software for its cars. But when it came time to rethink the telematic features for future models (such as GPS navigation), the company released a digital design kit on its Web site to encourage interested customers to design them instead. Thousands responded and shared ideas with company engineers, many of which have since turned into valued initiatives. Now BMW hosts a "virtual innovation agency" on its Web site, where small and medium-size businesses can submit ideas in hopes of establishing an ongoing relationship.

While decidedly less high-tech than BMW, John Fluevog is a designer of high-end shoes. He may not compete with Nike, but his world-famous shoes have been selling reliably to an expanding customer base since 1980. Inspired by the Linux phenomenon, Fluevog has created open source footwear (though the process only loosely resembles those employed in the open source software community). Customers submit designs for consideration, and the best ones get put into production. While Fluevog isn't offering royalties or placing the designs back into hands of "the community," he has promised to adorn any shoe design he adopts with the name of the designer.[6]

These cases illustrate how smart companies are reaching out to involve customers and lead users directly in their product development processes. One of the important elements not captured by these examples, however, is the extent to which customer innovation is going self-serve with the rise of prosumer communities.

Customer Co-Innovation Goes Self-Serve

David Pescovitz, senior editor for *Make* (a magazine and blog devoted to the do-it-yourself [DIY] innovation scene), says the DIY phenomenon is exploding with prosumer communities that have formed around products ranging from the Toyota Prius to the Apple iPod: "Communities are forming every day in part because the technology enables it." There is no need for users to innovate in isolation or wait for the next monthly amateur electronics meeting to share their customized wares. Pescovitz also highlights the allure of prestige and the sense of social belonging that develops within prosumer communities. "People get big thrills from hacking a product, making something unique, showing it to their friends, and having other people adopt their ideas," he said.

Even Hollywood is getting involved. The 2006 cult movie *Snakes on a Plane* engaged its audience in many aspects of the film ranging from scripting to marketing. Fans of star Samuel L. Jackson convinced the producers to insert lines into the dialogue and could create a custom and personalized voice message from Jackson to send to their friends. This inspired one blogger to proclaim that we're seeing a shift from "heard the ad, seen the movie, bought the video, got the T-shirt, got the fridge magnet"; to "created the ad (co-) shot the movie, mashed the video, designed the T-shirt, made the fridge magnet."[7]

One of the earliest, and still most vibrant, prosumer communities has formed around Lego products. Lego itself has become a flagship for how to get your customers deeply involved in cocreating and co-innovating products. Though Lego is perhaps best known for making little interlocking plastic bricks, the company is increasingly focusing on high-tech toys. With Lego Mindstorms, for example, users build real robots out of programmable bricks that can be turned into two-legged walking machines, or into just about anything a teenage mind can envision. When the product first made its debut in 1998, marketing officials were surprised to discover that the robotic toys were popular not only with teenagers but with adult hobbyists eager to improve on them.

Within three weeks of its release, user groups had sprung up and tinkerers had reverse engineered and reprogrammed the sensors, motors, and controller devices at the heart of the Mindstorms robotic system. When users sent their suggestions to Lego, the company initially threatened lawsuits. When users rebelled, Lego finally came around, and ultimately incorporated user ideas. It even wrote a "right to hack" into the Mindstorms software license, giving hobbyists explicit permission to let their imaginations run wild.

Today Lego uses mindstorms.lego.com to encourage tinkering with its software. The Web site offers a free, downloadable software development kit; Lego's customers in turn use the site to post descriptions of their Mindstorms creations—and the software code, programming instructions, and Lego parts that the devices require. Indeed, Mindstorms enthusiasts are notoriously ambitious. At Lego World 2005 in the Netherlands, one participant revealed a full-size, fully functional pinball machine made from twenty thousand Lego blocks and thirteen programmable microchips.

The company benefits hugely from the work of this volunteer business web. Each time a customer posts a new application for Mindstorms, the toy becomes more valuable. Lego senior vice president Mads Nipper calls it "a totally different business paradigm." "Although users don't get paid for it," he says, "they enhance the experience you can have with the basic Mindstorms set—it's a great way to make the product more exciting." We've always thought that Lego ought to make its most passionate devotees part of its design department. And when it came time to develop a new version of Mindstorms, NXT, in 2005, that's exactly what the company did, by taking four of its most prolific users on as de facto employees for the eleven-month development cycle.

The Mindstorms experience has proven to be so successful that Lego has transferred its customer-centric development practices to its more conventional Lego brick toys with a service that lets customers design their own custom Lego sets. Users no longer need adhere to the tyranny of Lego's predesigned kits. With the new Lego Factory system launched in 2005, customers get access to a virtual warehouse of Lego elements with which to design, share, and purchase custom models.

It's a bit like open source Lego: Simply download the free 3-D modeling program that lets you design your virtual toy, using as many bricks as you want. Upload your Mona Lisa to Lego's Web site and you—and any other Lego fan—can order the kit for your creation, complete with assembly instructions.

Sounds obvious when you think about it, and yet Lego's fusion of mass customization and peer production remains rare enough in today's consumer products market to make the idea particularly outstanding. Customers can make whatever they want, and Lego transforms its legion of youthful customers into a decentralized virtual design team that invents and swaps new Lego models. Mark Hansen, director of Lego Interactive Experiences, says, "With Lego Factory we can expand beyond our one hundred in-house product designers to marvel at the creativity of more than three hundred thousand designers worldwide." Indeed, with the Lego Factory and Mindstorms combined, the company has moved far beyond customer-centricity to harness a full-fledged prosumer community that will help ensure that Lego remains a vibrant source of innovation for many years to come.

THE PROSUMPTION DILEMMA: CONTROL VERSUS CUSTOMER HACKING

Prosumption sounds like a win-win proposition. Indeed, how could you possibly lose? Customers get more of what they want and companies get free R&D. But it's not all cut-and-dried. As prosumer communities proliferate, companies face increasingly tough choices about how to interact with them. Are customer innovations always good news? What happens when the modifications and extensions that customers develop conflict with a company's business imperatives? Should companies discourage, ignore, join, or even try to co-op prosumer communities? Lego has been fortunate. But for some companies these questions have become agonizing and perplexing.

Take Apple's iPod. The now ubiquitous music and media player is one of the most popular electronic devices to emerge in the last decade. Tens of millions of consumers around the world use the iconic device to bring their music and media with them wherever they go. It's been a tremendous success for Apple. Along with the complementary iTunes digital music service, the iPod has revitalized the company, while transforming both the music and consumer electronics industries single-handedly.

Perhaps not surprisingly in this day and age, Apple's customers are even more ambitious. Its lead users have always surmised that the iPod could be much more than a digital music player. The iPod, after all, is a brawny piece of hardware with a massive hard drive. Limiting it to playing music files would seem a shame, when so many other applications were possible. Why not transform the iPod into a general-purpose wearable computer that has everything from video games to Wikipedia?

All-purpose wearable computers may yet be in Apple's game plan (after all, Apple has partnered with Nike to integrate the iPod into the popular sports gear line). But the company is notoriously tight-lipped about its product road map, and is understandably riding the success of its music applications. Some users became impatient and endeavored to use the iPod as a platform for their own innovations. The problem for adventurous users is that the iPod is a closed system. There is no documentation for the software or tools to help developers turn it into something else. Of course, this has never stopped users before and, quite predictably, users have taken matters into their own hands, quite literally. Whether it's modifying the casing, installing custom software, or tearing it up and doubling the memory, users are transforming the ubiquitous music and media player into something unique. Tens of thousands of users gather at online forums to swap ideas and coordinate their actions. Of the hundreds of customer-inspired hacks that have emerged, the most powerful is a program called Podzilla—essentially a bare-bones version of Linux with a graphical user interface that runs on the iPod's tiny screens.[8]

Once users install the hack, they can either boot their iPod as usual, or fire up Podzilla for a pocket Linux environment. Podzilla utterly transforms the iPod, allowing users to view pictures, play several games, and record audio at full CD-quality by plugging in a microphone (Apple cripples the iPod so that it can record audio at 8kHz only). Add a keyboard that can be plugged into the headphone jack, and it could become a fully functional PDA, capable of editing calendars, address books, and e-mails. Arguably the project's most notable accomplishment is its DIY video player—released months before rumors about Apple's Video iPod had even begun to spread.

With Podzilla users get applications galore. In addition to enabling games such as Othello, Pong, Tetris, or Asteroids, hackers have reworked Doom so that it will play on the device, albeit at an agonizingly slow 3 to 4 frames per second frame rate. Another application, called PodQuest, allows you to download driving directions from Google Maps, MapQuest, Yahoo Maps, and others. Everybody loves Wikipedia. Now with Encyclopodia you can get it on your iPod and carry Wikipedia with you everywhere you go. Bold hackers have even figured out how to double the 4-gigabyte memory in the stingy iPod Nano. Just buy a broken Nano from eBay, pop out its memory chip, solder it into the empty slot in your working Nano, and hit reboot—this hack is for advanced users only!

So far Apple has stayed largely silent on its customers' transgressions—they don't explicitly condemn product hacking, but they don't condone it either. Apple has refused to release a developer kit that would make it both legitimate and easy for users to modify or build on the iPod platform. But Apple CEO Steve Jobs has yet to unleash his lawyers or publicly denounce his customers.

Jobs knows the company walks a very fine line. Apple's iTunes/iPod business model is built on its very lack of interoperability with other devices and services. For example, Apple's digital rights management software—euphemistically called FairPlay—prevents consumers from making unlimited copies of iTunes songs and ensures that the iPod doesn't work with any other copy-protected formats. This means customers are forced to buy their music through iTunes. Likewise, competitors like Real Networks can't legitimately sell music through their own online services that will play on Apple's iPod. As Steve Jobs himself said: "With iTunes, we decided to work with the most popular music player—and that's by far the iPod. Rather than support all these other guys, we'd rather use the engineering to innovate."

But what happens when "the other guys" are not just competitors but your most loyal and engaged customers? The iPod's closed architecture is good at keeping competitors at bay, but it also limits what users can do with the device. That may prop up Apple's business model. It may even allow Apple to add new features and capabilities incrementally in order to keep customers coming back for more. But is a business model that locks in customers and discourages user innovation genuinely sustainable?

Only Steve Jobs knows for sure where Apple wants to take the iPod next. The company has already entered the portable video and phone markets in a big way. As Apple plans its next move, there is little doubt that the company is watching its lead user communities closely and taking cues from what they do with the device.

At the same time, Apple executives must worry that if users can reengineer the product to add a seemingly unlimited array of new features and capabilities, there will be little incentive for customers to spend more money at the Apple store to upgrade to new iPod versions. Any move to open up the iPod's closed architecture ends up threatening both the viability of its current business model and Apple's future product strategy.

Customer Hacking and Home-Brew Applications

Apple is not alone in its muddling efforts to figure out how to deal with increasingly savvy, impatient, youthful, and technically sophisticated customers who insist on taking their technology to the limits. Sony's popular PlayStation Portable (PSP) has also become a platform for a wide range of customer hacks that precociously extend the capabilities of the portable video game player.

Like iPod and Lego Mindstorms enthusiasts, Sony's customers were quick to rip into the system. Within days of hitting the shelves, PSP fanatics were adding new unauthorized capabilities and features. Now PSP customers go online in vast numbers to swap home-brew applications and games on a variety of user-developed Web sites. Some of the more ingenious user-engineered hacks have turned the PSP into a streaming music player, a WiFi device, and a Web browser. Even relative novices can enjoy these clever extensions by following carefully prepared instructions.

Sony goes further than Apple in explicitly denouncing its customers' ingenuity. The company has even taken steps to retroactively lock up its PSP platform. Before users can load Sony's latest games and peripherals, for example, they must upgrade the PSP's firmware (the operating software that runs the PSP). Frustrated customers learn after the fact that Sony's new firmware disables all of the home-brew games and applications that they worked hard to develop on previous versions. Inevitably it has been a losing battle—hackers crack the new firmware versions just as fast as Sony can release them. But when questioned by the media about why the company repeatedly cripples features that make the PSP more attractive to customers, a Sony rep could only stutter: "Consumers should be aware that any hacking or home-brew applications may cause damage to the PSP unit and may void the warranty."

Of course, Sony's war on product hackers has little to do with warranties and much more to do with its business model. Like Apple, Sony's business model is tied not just to device sales, but to complementary sales of PSP content and peripherals—notably, in Sony's case, the lucrative gaming market for its console. Allowing users to develop their own sources of entertainment for the PSP is tantamount to cannibalizing its offering. And, like Apple, Sony fears it could lose control of its platforms and perhaps even create opportunities for new competitors.

Embracing Consumer Power

So here's the prosumption dilemma: A company that gives its customers free reign to hack risks cannibalizing its business model and losing control of its platform. A company that fights its users soils its reputation and shuts out a potentially valuable source of innovation. Apple and Sony may feel the latter option is an acceptable risk so long as hacking remains at the fringes. After all, product hackers are still a small minority of their customers, and there is little evidence yet that product hacks and home-brew applications are leaking out into the mainstream. Any company that believes that the status quo will hold for long, however, is mistaken. Product hacking is just getting started.

Customers with the skills and inclination to hack their products may be in the minority today, but what about in five or ten years, as increasingly technology-savvy kids become the norm? Will companies choose to fight all of their customers then? How will they cope with the proliferation of tools and Web sites that enable prosumer communities to flourish? Will they unleash the lawyers and risk driving their customers to alternative platforms? Indeed, how will they compete with the inevitable rise of hacker-friendly platforms that let customers do whatever they want and in return tap unlimited pools of free innovation? The answer is they can't and won't fight their customers for long. Customer hacking will live on.

Smart companies will bring customers into their business webs and give them lead roles in developing next-generation products and services. This may mean adjusting business models and revamping internal processes to enable better collaboration with users. It certainly means avoiding Sony's practice of disabling customer innovations. That is a small price to pay, however, to keep customers loyal to your business.[9] In fact, the opportunity to generate vibrant customer ecosystems where users help advance, implement, and even market new product features represents a largely untapped frontier for farsighted companies to exploit. We will return to flesh out some of these ideas in the conclusion to this chapter. For now, we turn to the rise of listener-artists and the Cambrian explosion of creativity on the Web.

LISTENER-ARTISTS AND THE CAMBRIAN EXPLOSION OF CREATIVITY

Lego hobbyists and iPod hackers give us a taste of this new prosumer ethic, and of both the challenges and opportunities it raises in various industries. Perhaps the most exciting and broadest frontier of user creativity, however, is happening on the Web where amateur artwork, music, photos, stories, and videos comprise an explosion of cultural innovation that is flowing through blogs, wikis, podcasts, Internet television sites, and a variety of peer-to-peer distribution channels.

This rich, diverse outpouring of creativity is driven by a convergence of peer-to-peer networks, inexpensive digital devices, open source software, user-friendly editing tools, cheap storage, and reasonably affordable bandwidth. The result is that users can create and share content to amuse themselves; individuals with a point of view can influence the media agenda; and community sites with advertising can cut deeply into revenue that normally would go to media conglomerates.

This has put media companies at odds with their customers. Indeed, in no other industry is the tension between the preexisting power of producers and the increasing power of self-organized customer communities so pronounced. Nothing illustrates the opportunities and trade-offs of prosumption better than the growing propensity of young people to weave fluid and participatory tapestries of music content into their own unique and inviting creations. Call it "the remix culture."

Remix Culture

Lawrence Lessig likes to remind people that cultural remixing is nothing new. "Since time immemorial people have been engaged in the act of remixing their culture," says Lessig. "They would do it in obvious simple ways like watch a movie and retell the story to their friends, or they would use a sitcom as a basis for a cultural reference or a joke, but the point is that they are constantly using this culture in their ordinary life and sharing it with others in day-to-day conversation."

Of course, the difference today is that technology makes it easy for people to remix culture and share it on a much larger scale. Not only can people share their remixes with three or four best friends, they can now share them with thousands, and perhaps millions, on the Web.

Though Hollywood argues otherwise, remixing music is not about copying artistic works; it's about modifying, embellishing, appending, reinventing, and mashing them together with other elements. Most of all, remixing music is about being a producer, participating in the creative enterprise, and sharing your creations with others. "That is what digital technology is doing," says Lessig. "It is infinitely expanding the technological capacity for participation in this kind of creative work." Even the great Italian Renaissance of the fifteenth century will pale by comparison if these creative energies are allowed to flower.

Where did this remix culture come from? Its modern incarnation arguably begins with hip-hop. Starting in the early 1970s, hip-hop artists began mixing and matching beats from various sources, and then layering their own rhythmic vocals on top. This new art form proved highly popular with young people, and now constitutes one of the industry's most lucrative genres.

Despite, or perhaps because of, its growing popularity, this fresh approach to making music attracted its fair share of critics—not least of which were the artists and record companies who didn't like hip-hop acts "sampling" their work. As Public Enemy producer Hank Shocklee recently explained, "We were taking a horn hit here, a guitar riff there, we might take a little speech, a kicking snare from somewhere else. It was all bits and pieces."[10]

Hip-hop artists claimed fair use, while record companies cried that two-second samples of a catchy rhythm, melody, or sound infringed their copyrights. The record companies won in court, and today samples of any length or description (not just recognizable samples) have to be cleared legally with copyright owners before a song or album can be released.

Many in the industry fear that legal encumbrances are chilling musical innovation. In hip-hop's heyday innovative producers literally layered hundreds of samples and snippets to create a collage of sound fashioned into a new song. Today the cost of clearing samples and producing albums is rising so quickly that the most creative works will never be heard.[11]

Bedroom DJs

But like most forms of popular culture that encounter official condemnation, hip-hop continues to grow more popular, and its derivatives are popping up in new places all the time. In fact, as the software to manipulate and remix music proliferates, hundreds of bedroom DJs and songwriters have emerged to make their own "bastard pop" confections. "You don't need a distributor," says Mark Vidler, known professionally as Go Home Productions, "because your distribution is the Internet." "You don't need a record label," he continues, "because it's your bedroom, and you don't need a recording studio, because that's your computer. You do it all yourself."[12]

The most popular form of DIY creativity is what participants call "mashups," "bootlegging," "bastard pop," and a variety of other labels. The common theme is that aspiring artists fuse songs digitally from completely different genres to produce hybrid singles and, increasingly, full-length mashup albums.

Want a new twist on your well-worn Beatles collection? Try DJ Danger Mouse's *Grey Album*, which consists entirely of contorted samples from the Beatles' *White Album* mashed together with vocals from Jay-Z's smash hit *The Black Album*. Or how about bacchanalian rapper Missy Elliott combined with the gloomy melodies of English rock band Joy Division, or Madonna's elated voice layered over a grinding Sex Pistols track.

Unorthodox? Yes. Illegal piracy? Perhaps. Innovative and enjoyable? Most definitely. In fact, a growing number of music lovers are convinced that this is the future of participatory music. Even music critics agree that many mashups more than exceed the sum of their parts. And as the phenomenon has slowly gained acceptance, large communities of mashup makers have been coming out of the shadows. They congregate on the Web in growing numbers, where they eagerly offer critiques of new songs, tips for newbies, pointers on where to find a cappellas, legal advice, publicity for mashup events, and general discussion of issues surrounding the mashup phenomenon.

Now, if the record companies would only wake up to the opportunity, they would be falling over backward to create platforms to encourage creative remixing—perhaps offering subscriptions for access to the best tools and tracks. But like hip-hop artists before them, lawyers have gone after mashup artists like a pack of rabid dogs.

"The problem," says Lawrence Lessig, "is that according to copyright law mashups are illegal, and increasingly, as record labels learn about mashup artists they are doing what their lawyers say that they have to do, which is to stop it and shut it down." Mashup artists typically spend an extraordinary amount of time producing extremely creative stuff that has one effect, and that is to promote the underlying music. Though the original artists, the fans, the creators, and ultimately the labels stand to benefit, the labels won't sanction it unless they themselves control it. "The people producing mashups are furious," says Lessig, "and the mashups themselves no longer promote the work of the artist. Yet the existing regime of copyright says that this is absolutely obvious, that this is what you should do, and the claim of 'they have a right to do this' turns out to be very, very weak."

The Latter-Day Aristocracy of Creativity

The labels' logic is flawed. When Danger Mouse's *Grey Album* was released on the Web it was an overnight sensation. But within weeks of its catching on, EMI had issued cease-and-desist letters to every Internet distributor it could uncover. But as blogger and copyright activist Cory Doctorow put it, "No one who listens to *Grey Album* will shrug her shoulders and say, 'Well, heck, now that I've heard that, who needs to buy the Beatles album, or Jay-Z's album?' " On the contrary, it makes the Beatles and Jay-Z more popular by exposing their work to new audiences. Ironically, Danger Mouse was later hired by EMI to produce mashups legally for the company.

But this is hardly a coup for mashup artists. Doctorow points out that copyright lawyers like to contrast copyright with the old system of patronage, when you could only make art if you could convince the pope or a duke or a king that your art was worthy. Patronage distorted creative expression, and copyright did much to decentralize authority over artistic acts. Part of what makes mashups great—and what makes the revolution in user creativity on the Web important generally—is the completely decentralized, spontaneous, and unimpeded way in which new content is produced. At least until the recording industry started cracking down, creators felt free to let expression, creativity, and audience responses—not legal concerns—guide their creations, just like the early days of hip-hop and other user-controlled frontiers.

EMI's answer to the *Grey Album* sounds a lot like the old patronage system. "If you work for one of a few big record companies," says Doctorow, "you can use their legal apparatus to clear the material you want to use in a mashup. Otherwise, your art is illegal."

The ability to continue to produce art without permission from the latter-day aristocracy of creativity is central to both cultural and economic progress. Whether it's bedroom DJs, garage innovators, or scientists in an advanced research laboratory, we don't want them to be consulting with attorneys all of the time about the legality of what they're doing. Nor do we want them asking technologists for the encryption keys before they can even begin to engage in an act of creative enterprise. So much of what makes a free society and free economy healthy and vibrant is that we have limited the control points in a way that permits creation and experimentation in a largely anarchistic fashion.

The Open Hand

Fortunately, in the case of copyright there are alternatives, but it's no surprise that they come from grassroots movements and not the recording industry. Consider the Creative Commons, an initiative launched in 2002 that offers content creators flexible licenses for managing their creative rights. For most artistic types, licensing work can be a nightmare, and an expensive one at that. The basic default rule of copyright, says Creative Commons architect Lawrence Lessig, is that all rights are reserved, and the basic infrastructure of copyright is, "Talk to my lawyer if you want something less than that." That means that the cost of actually negotiating around the default is very high.

Creative Commons (www.creativecommons.org) provides licenses that allow you to protect your copyright ownership while allowing others to make derivative works, and stipulating whether you only want to allow non-commercial or commercial use, among many other options. If you have an audio track you'd like to let other people post freely or sample, for example, just affix a CC license, and the world is now free to use it. A growing number of artists, writers, musicians, photographers, and other creators are seeing the benefits of this more flexible and hassle-free option.[13]

The Creative Commons has even spawned a new mashup platform called "ccmixter.org" where participants can remix CC-licensed content and share it with the community. "It's a community of people that just couldn't exist if it weren't for this kind of licensing," says Lessig. "In the past, if you put all this material on the Web and said, 'Hey, come remix this stuff,' it would take a week before you got a cease-and-desist notice from the RIAA [Recording Industry Association of America]. So this is effectively creating a possibility for a kind of creativity that otherwise would just not have been allowed."

It's not just amateur artists who are getting in on the action. Major artists such as David Byrne, the Beastie Boys, Nine Inch Nails, and many others are getting involved. These bands see fan-created remixes as a way to connect with their audience, and they in turn help bands extend their reach and musical repertoire.

The Beastie Boys, for example, have been posting a cappella versions of their songs and encouraging fans to mash them up with their own music tracks for years. Remixes can be used for noncommercial purposes, and are available for download on www.beastieboys.com.

In October 2004, the band took it a step further and decided to get their fans to help make a documentary film of an upcoming concert. So the group recruited fifty fans on the Internet, equipped them with Hi8 video cameras, and set them loose in Madison Square Garden. The fans' only instructions for documenting the concert: Start filming when the Beastie Boys hit the stage and don't stop filming until it's over.

The end product—a kaleidoscopic collage of amateur video called *Awesome: I Fuckin' Shot That!*—was cobbled together by band member Adam Yauch (aka MCA or Nathaniel Hornblower) from over sixty angles and one hundred hours of footage.[14] The film's coproducer, Jon Doran, calls it "the democratization of filmmaking."

Of course, democratization is a scary word for those accustomed to ironclad control over the creation and distribution of music. "But at some point," says Jim Griffin, the former head of technology at Geffen Records, "the music industry must come to a realization that they can hold a great deal more in an open hand than they can in a closed fist."

Digital music provides the occasion for this realization. It offers a historic opportunity to place artists and consumers at the center of a vast web of value creation. But these novel dynamics have turned the record industry on its head. Rather than build bold new business models around digital entertainment the industry has built a business model around suing its customers. With artists now increasingly turning against the record industry's lawsuits, however, momentum may be shifting in favor of a better way forward.

In fact, the music industry saga serves to illustrate a fundamental principle: Customer value, not control, is the answer in the digital economy. The music industry—and all industries for that matter—must resist the temptation to impose their will on consumers as a matter of convenience, or worse, as a result of a lack of ingenuity and agility. Rather, music labels should develop Internet business models and offerings with the right combination of "free" goods, consumer control, versioning, and ancillary products and services. This includes new platforms for fan remixes and other forms of customer participation in music creation and distribution.

WE ARE THE MEDIA

The rise of citizen journalism and consumer-controlled media provides yet another example of how mass collaboration and cocreation are erasing the previous boundaries between companies and consumers. In a world where all one needs is a camera phone to report on one's surroundings, it is no longer as straightforward to pigeonhole a person's role. In the emerging prosumption paradigm, a person can seamlessly shift from consumer to contributor and creator. Consider these examples.

YouTube is the latest in a string of Internet-TV offerings that makes it ludicrously easy to publish, play, and share video clips on the Web. Anyone can upload a video to the site, and millions of members relish the opportunity to heap praise on the clever videos, while the less clever get seriously flamed. Really popular videos spread with viral intensity, attracting millions of viewers who clamor to see what all the hype is about. It all comes together in a slightly anarchic and unceremonious fashion. But with a global audience providing all of the programming, scheduling, and commentary, the experience of just browsing YouTube is novel entertainment in itself.

At the time of writing, YouTube was offering a motley collection of home movies, independent films, and pirated video content. Users can see everything ranging from clips of their favorite soccer players to U.S. soldiers capturing scenes of combat in Iraq. Though much of the original content is amateurish, it can make for surprisingly captivating viewing. Of course, the Net Geners are all over it, and many use it to share their home videos with friends (or anyone else who's interested). Budding deals with Hollywood could make the service even more popular, and turn YouTube into a major distribution hub. With over one hundred million plays a day and growing, it seems likely to be a force to reckon with.

Another early and important example of how consumers are re-defining the media experience is Slashdot, where thousands of people upload news items of interest to a global audience of techies and pro-grammers. The value of any particular news item is determined by the ratings of readers and moderators on the site. Site traffic is so great that the term "to be slashdotted" has entered the lexicon, meaning your own site got an overwhelming amount of hits based on a single mention on Slashdot.

If Slashdot is the grandfather of reader-compiled technology news sites, digg is the prodigy child. It's a lot like Slashdot, except digg is more egalitarian. Slashdot has a top-down editorial structure. Only editors can select user-submitted news items to display on the home page. Visitors can't see all the stories users submit. Nor can they vote on them.

By contrast, digg is refreshingly simple and democratic. Members rec-ommend interesting stories to one another by posting links to the digg site. Healthy competition to discover great stories makes digg a vibrant source for timely tech news. "The members get credit for being the first to find stories," says Jeff Jarvis, a media consultant, blogger, and avid digg participant, "which means that you have over 150,000 editors competing to find the good stuff fast."[15]

Once the articles appear on digg, members click to check them out, sending a tsunami of traffic to each article. One link on the front page can cripple a server for days. It's just like being "slashdotted," except members call this "the digg effect."

Users exert editorial control by clicking on the digg button for each story they like. Articles that receive the most diggs are promoted to the home page. And so the community collectively updates the front page. You could say the community is the editor.

Jarvis says his fourteen-year-old son is also "addicted to digg." Jarvis thinks that's great. "It proves that young people do care about news," he says. "You can go to digg.com/spy and watch the public swarming around stories they like," says Jarvis. "My son can see the stories his friends like.

I can subscribe to a feed of the stories he likes. The news is a community activity again."[16]

Why is this happening on digg and not CNN or the New York Times? It's simple. Digg's creators have learned how to make news a social pastime. And, like all other facets of their lives, prosumers want in on the conversation.

Techies like to squabble over whether Slashdot or digg has the better model. Slashdot is renowned for quality and highly technical discussions. Digg is known for its immediacy and the sheer volume of aggregated stories (thousands every day). Regardless of their differences, both sites make most traditional news outlets look like archaic relics of a bygone era, especially when it comes to the way these sites interact with and relate to their audiences.

Perhaps it's because the mainstream media just doesn't get it yet. Jarvis says mainstream news editors look at sites like digg and worry that second-rate stories will make it onto the front page. But are editors really in a position to best the collective judgment of their audience? Maybe they're worried that it will go a step further. We'll let journalists post their stories directly and let the community decide which stories are newsworthy and important. After all, if the community is the best arbiter of relevance, do we really need editors?

In truth, serious news organizations will always require great reporters, writers, and editors to deliver top-notch content. Above all, they need individuals with the skills and experience to ferret out great stories and editors with the authority to uphold standards of independence, professionalism, and accuracy. Digg and Slashdot have the easy job by comparison—they aggregate, rate, and comment on the news, they don't do the hard-core reporting.

Democratizing the Media

Nevertheless, there is a lot to learn from these examples. If mainstream outlets were to engage and cocreate with their audiences in a more profound way, surely this could only accentuate positive attributes such as balance, fairness, and accuracy, while making the media experience more dynamic.

For example, any serious news organization today should also allow its community of readers to join in the editorial conversation. The fact that all major media properties don't already offer a parallel front page edited by readers is troubling. The technology has been available for a decade. A cynic might call it contempt for the collective intelligence of media consumers. In some cases, the cynics may be right. But in most cases the sclerotic pace of change reflects the cultural inertia of institutions steeped in the journalistic traditions of mass media.

The new Web challenges the assumption that information must move from credentialed producers to passive consumers. "The mainstream media people define themselves as the arbiters of taste," says Judy Rebick, founder of Rabble, a thriving community-driven news media and discussion forum in Canada. "As long as the media thinks they know what's right," she continues, "they'll never be in a position to harness people's collective intelligence. It's a completely different culture and a completely different way of thinking about knowledge."

The democratization of the media publishing tools, however, is rapidly transforming our notions of how expertise, relevance, and professionalism develop in the media. "The old way of thinking," says Rebick, "is that the cream rises to the top. . . . [Y]ou have hierarchical structures that cut people out at each level." On sites like Rabble, the users, not managers, make those decisions. "Instead of cutting people out, we bring them in, and people can pick and choose what they want to read or hear. They don't have to listen to all the podcasts or read all of the blog posts. But there's something there for everybody, and it allows for people to come in and do their thing and get noticed."

There are small signs that the mainstream media is changing. A recent poll asked media executives for ideas as to how big media firms could respond to the new "threats" posed by user-generated content. Their responses read like a prosumption playbook.

Suggestions included:

- Give users access to raw content such as interviews as a means of providing greater transparency and accountability.
- Provide tools and become a platform for user-generated rather than firm-generated content.

- Redesign all content to be a conversation rather than a corporate monologue.
- Treat advertising as content too.
- Use new distribution forms, including peer-to-peer networks.
- Adapt content forms and schedules to user demands.

Actions speak louder than words, however, and few of these ideas have been championed. A continued lack of responsiveness will be their ultimate downfall. Media organizations that fail to see the writing on the wall will be bypassed by a new generation of media-savvy prosumers who increasingly trust the insights of their peers over the authority of CNN or the *Wall Street Journal.*

HARNESSING PROSUMER COMMUNITIES

Prosumption is becoming one of the most powerful engines of change and innovation that the business world has ever seen. Cocreating with customers is like tapping the most uniquely qualified pool of intellectual capital ever assembled, a reservoir of talent that is as keenly and uniquely enthusiastic about creating a great product or service as you are. But it comes with new rules of engagement and tough challenges to existing business models. Anyone who tells you different has not fully grasped the implications of the impending prosumer revolution.

More than customization

Just as prosumption is more than marketing disguised as customer advocacy, it goes way beyond product customization. Customization occurs when a customer gets an off-the-shelf product adjusted to his or her specification. There is nothing wrong with mass customization: Customers get to tailor products to specific uses while companies get to maintain the economies of large-scale production.

The problem is that mass customization generally entails mixing and matching prespecified components, which significantly limits flexibility and innovation for users. When you order a Dell computer, for instance, you can slot in any DVD drive you want, but it's still a DVD drive. True prosumption entails deeper and earlier engagement in design processes (i.e., Lego's next-generation Mindstorms) and products that facilitate customer hacking and remixing (mashups).

Losing control

Customers will increasingly treat your product as a platform for their own innovations, whether you grant them permission to or not. As both the iPod and PSP cases illustrate, they invent new ways to create extra value by collaborating and sharing information. Over time value migrates from your product or service to what customers do with the information. If you do not stay current with customers, they invent around you, creating opportunities for competitors. Inevitably, it is preferable to sacrifice some control than it is to cede the game completely to a more adept, prosumer-friendly competitor.

Customer tool kits and context orchestration

Forget about static, immovable products. If your customers are going to treat products as platforms anyway, then you may as well get ahead of the game. Make your products modular, reconfigurable, and editable. Set the context for customer innovation and collaboration. Provide venues. Build user-friendly customer tool kits. Supply the raw materials that customers need to add value to your product. Make it easy to remix and share. We call this designing for prosumption.

Becoming a peer

After gaining some experience with this new world of prosumption you'll realize that your real business is not creating finished products but innovation ecosystems. Companies will participate in these ecosystems in the same way that IBM participates in open source—it harvests value from Linux, but it does not own or control the Linux ecosystem. Similarly, Second Life creates an environment in which customers do 99 percent of value creation. As prosumption matures, expect to treat customers like peers, not patrons.

Sharing the fruits

Customers will expect to share in the ownership and fruits of their creations. If you make it profitable for customers to get involved, you will always be able to count on a dynamic and fertile ecosystem for growth and innovation.

Don't think communism. Think of the eBay microeconomy instead. Hundreds of thousands of eBay's customers make their living there, while eBay takes a cut of their transactions. Indeed, with Second Life's customers creating so much of the game content, it only seems right that they should own all of the IP rights to their creations and make real money by selling in-game assets. IP rights spur prolific rates of customer cocreation and make Second Life's thriving virtual economy a source of real-world income for customers. Why couldn't your products and services support similar kinds of value-added activities?

The Future of Prosumption

The old customer cocreation idea was simple: Collaborate with your customers to create or customize goods, services, and experiences, all while generating a built-in market for your wares. Listen to your customers and run design contests or other such promotional schemes—basically anything that will get your most loyal and engaged customers to share their intellectual capital for free. In exchange, customers with the best ideas get a direct say in what actually gets produced. Maybe, if they are lucky, they get a small cash or in-kind bonus too.

This is the company-centric view of cocreation. *We'll* set the parameters by telling you when and on which products to innovate. *You'll* give us your ideas for free, but *we'll* choose the best of them—and keep all of the rewards and IP. Sound like a good deal?

Let's just say that most customers—especially those of the Net Generation—don't think so. In the new prosumer-centric paradigm, customers want a genuine role in designing the products of the future. It's just that they will do it on their own terms, in their own networks, and for their own ends. In fact, they will do so increasingly without you even knowing about it. Products that don't enable and invite customer participation will be anathema—staid, old-fashioned remnants of a less customer-friendly era.

If you expect to be around in the next decade, your organization will need to find ways to join and lead prosumer communities. Just remember: Customers won't care whether their activities make *you* more money (that's your job)—they'll just want a superior product and experience, and perhaps even a cut of the revenue. But just as IBM and other tech firms create billions of dollars of revenue by collaborating with the open source community, consumer product companies can find ways to monetize customer-led ecosystems.

And think about the possibilities for you as an individual. You're no longer just a passive recipient of products and services. You can participate in the economy as an equal, cocreating value with your peers and favorite companies to meet your very personal needs, to engage in fulfilling communities, to change the world, or just to have fun! Prosumption comes full circle.

6. THE NEW ALEXANDRIANS
Sharing for Science and the Science of Sharing

The Alexandrian Greeks were inspired by a simple but powerful idea. Collect all of the books, all of the histories, all of the great literature, all of the plays, all of the mathematical and scientific treatises of the age and store them in one building. In other words, take the sum of mankind's knowledge and share it for the betterment of science, the arts, wealth, and the economy. The Alexandrians came very, very close to achieving that goal. At its crowning glory, estimates suggest that they had accumulated more than half a million volumes.

Certainly the works of great thinkers such as Aristotle, Plato, and Socrates could be found there. It was also the place where Archimedes invented the screw-shaped water pump, Eratosthenes measured the diameter of the earth, and Euclid discovered the rules of geometry. Ptolemy wrote the Almagest at Alexandria; it was the most influential scientific book about the nature of the universe for the better part of 1,500 years. And for those reasons the Great Library of Alexandria is regarded by many as the world's first major seat of learning, perhaps even the first university, and the birthplace of modern science.

When the library was destroyed in the fifth century it was a major setback for the arts and sciences. Five hundred years later, the largest library had less than one thousand volumes. With forty-two million items today the New York Public Library is larger than the Alexandria library, but there are still very few libraries that rival the collection at Alexandria nearly two thousand years ago. This despite the fact that the stock of human knowledge is now infinitely wealthier than it was in the fifth century.

Indeed, we are fortunate to be living through the fastest and broadest accumulation of human knowledge and culture ever. *Wired* cofounder

Kevin Kelly recently reported that humans have "published" at least 32 million books; 750 million articles and essays; 25 million songs; 500 million images; 500,000 movies; 3 million videos, TV shows, and short films; and 100 billion public Web pages—and most of this knowledge explosion took place in the last half century.[1] Now add the constant stream of new knowledge created every day; so much, in fact, that the stock of human knowledge now doubles every five years.

Thanks to a new generation of Alexandrians this fountain of knowledge, past and present, will soon be accessible in ways our ancestors could only dream of. Companies such as Google, and librarians at esteemed institutions such as Harvard, Oxford, and Stanford, are hastily scanning books by the thousands and turning them into bits. Along with media of all varieties, these digitized books will be sewn together into a universal library of knowledge and human culture. When the new virtual library of Alexandria comes to fruition it will provide a shared foundation for collaboration, learning, and innovation that will make the present Internet look like a secondhand bookshop.

Digital libraries, and the Herculean efforts to build them, are impressive and important. However, they are only one aspect of a much deeper transformation in science and invention that we describe in this chapter. Indeed, the Alexandrian revolution extends far beyond the way we archive knowledge, to the way we create and harness knowledge to drive economic and technological progress.

A new age of collaborative science is emerging that will accelerate scientific discovery and learning. The emergence of open-access publishing and new Web services will place infinite reams of knowledge in the hands of individuals and help weave globally distributed communities of peers. The rise of large-scale collaborations in domains such as earth sciences and biology, meanwhile, will help scientific communities launch an unprecedented attack on problems such as global warming and HIV/AIDS. All considered, leading scientific observers expect more change in the next fifty years of science than in the last four hundred years of inquiry.[2]

As new forms of mass collaboration take root in the scientific community, smart companies have an opportunity to completely rethink how they do science, and even how they compete. Companies can scale and speed up their early-stage R&D activities dramatically, for example, by collaborating with scientific communities to aggregate and analyze precompetitive knowledge in the public domain. In fact, the efforts described in this chapter, including the SNP Consortium and Intel's open university network, suggest that even competitive rivals are seeing the benefits of collaborating on initiatives that will establish and grow a market for new products and services. Depending on the type of venture, firms can identify and act on discoveries more quickly, focus on their area of competence, facilitate mutual learning, and spread the costs and risks of research.

If this plays out the way we predict, the new scientific paradigm holds a more than modest potential to improve human health rapidly, turn the tide on environmental damage, advance human culture, develop breakthrough technologies, and explore outer space—not to mention help companies grow wealth for shareholders. That's a bold statement. But there is growing evidence to support it. Companies and scientific communities can harness mass collaboration to fundamentally change the world we live in. Read on to find out how.

THE SCIENCE OF SHARING

Humanity's capacity to generate new ideas and knowledge is the source of art, science, innovation, and economic development. Without it, individuals, industries, and societies stagnate.

In the past, firms have relied heavily on closed, hierarchical approaches to producing and harnessing knowledge. Increasingly, though, knowledge is the product of networked people and organizations looking for new solutions to specific problems. This peer-oriented approach to producing knowledge and sharing information is nothing new in academia: Research in the sciences has been circulating and building on discoveries for centuries. But it's new territory for firms.

Collaboration, publication, peer review, and exchange of precompetitive information are now becoming keys to success in the knowledge-based economy. As we have explained in previous chapters, the driving force behind this shift is the digitization of information and communications. Whether we look at art, science, commerce, or culture, we see that these forces are changing the way value is created throughout society. Digitization means information can be shared, cross-referenced, and repurposed like never before. Knowledge can build more quickly within networks of firms and institutions that cross seamlessly over disciplinary boundaries.

Conventional economic wisdom says companies should hoard their knowledge and technology. Most companies get prickly when people outside the firm start sharing or remixing their intellectual property. "I can't move into your backyard and just decide what to do with your landscaping," says Carla Michelotti, senior vice president and general counsel of Leo Burnett. "It's trespassing. It's taking somebody else's property."[3]

But in today's networked economy, proprietary knowledge creates a vacuum. Companies that don't share are finding themselves ever more isolated—bypassed by the networks that are sharing, adapting, and updating knowledge to create value. Conversely, evidence is mounting that sharing and collaborating, if done right, creates opportunities to hitch a ride on public goods and lift all boats in the industry. But first we must recognize that the modes of interaction in science (i.e., openness, peering, and sharing) have commercial viability, productive capability, and the ability to be drivers for private companies.

This is a considerable leap of faith for many managers who think the realm of science and the world of private enterprise operate under completely different principles. But it's not such a stretch when we recognize that just like science, the creative engine of capitalism requires access to the ideas, learning, and culture of others past and present. Indeed, the history of capitalism is replete with examples of how the material success we enjoy today is directly attributable to the evolution of openness in science and private enterprise and the rapid technological progress this unleashed. To see how this works in practice, it's worth taking a short detour into the history books.[4]

The Industrial Enlightenment

Starting as early as the seventeenth century—as the ideas of the Enlightenment took hold—we began to create, accumulate, and harness knowledge in new ways.[5] Engineers, mechanics, chemists, physicians, and natural philosophers formed circles in which access to knowledge was the primary objective. They exchanged letters, met in Masonic lodges, attended coffeehouse lectures, and debated in scientific academies. Some of these personal exchanges were confined to the realm of science. But a growing number helped smooth the path of knowledge from scientists and engineers to those who applied this knowledge to solve practical problems and build new cottage industries.

Buoyed by improved literacy rates, universal schooling, and the invention of movable type—and paired with the appreciation that such knowledge could be the base of ever-expanding productivity and prosperity—these nascent knowledge networks soon became indispensable to technological progress. For the first time in history, knowledge about the natural world became increasingly nonproprietary. Scientific advances were shared freely within informal scholarly communities and with the public at large. Science became a public good rather than the exclusive property of a privileged few.

The knowledge revolution continued into the eighteenth and nineteenth centuries, driving not just new knowledge and new ideals, but also better and cheaper access to knowledge and scientific tools. Improvements in our ability to publish and distribute knowledge, for example, dramatically lowered access costs, particularly for rank-and-file practitioners. It made the process of learning and economic change more efficient. Superior techniques spread faster. New technologies were more widely deployed and improved. More minds were trained in science, and more skills were brought to bear on practical problems.[6]

As time went on, the interplay between open science and private enterprise initiated a virtuous circle of knowledge creation and application that unleashed a period of sustained growth, prosperity, and technical improvement. Feedback from knowledge to technology—and from technology back to knowledge—made the continued evolution of science and learning the norm rather than the exception.

Eventually the wholesale pursuit of science radically improved our understanding of the natural world and enabled us to manipulate nature in previously unimaginable ways. Corporations arose as a vehicle to channel investment capital into entities capable of harnessing this new knowledge by turning it into products and services that the market desired. Further improvements in information technologies meant that knowledge was readily available for others to build on and improve. Over time, we honed the positive feedback loops between science and private enterprise to provide a sustained basis for economic growth.

The Age of Collaborative Science

Though the industrial enlightenment gave us much to be thankful for, it's fair to say that we still haven't seen anything yet. The advances in our capacity to generate and apply new knowledge in the industrial era pale in comparison to capabilities within reach today. Plummeting computing and collaboration costs are widening the distribution of knowledge and power. At the same time, our ability to self-organize into large-scale networks is enhancing our ability to find, retrieve, sort, evaluate, and filter the wealth of human knowledge and, of course, to continue to enlarge and improve it.

That's where we stand today. But as we spoke to colleagues and people out in the field we became convinced that we are only at the beginning of an exciting new scientific paradigm—call it the age of collaborative science. Just as the Enlightenment ushered in a new organizational model of knowledge creation, the new Web is helping to transform the realm of science into an increasingly open and collaborative endeavor characterized by:

- the rapid diffusion of best-practice techniques and standards;
- the stimulation of new technological hybrids and recombinations;
- the availability of "just-in-time" expertise and increasingly powerful tools for conducting research;
- faster positive feedback cycles from public knowledge to private enterprise, enabled by more nimble industry-university networks; and
- increasingly horizontal and distributed models of research and innovation, including greater openness of scientific knowledge, tools, and networks.

Above all, the new scientific paradigm will be truly global, swelled by the participation of millions of budding scientists from across Asia, South America, and Eastern Europe.

These are the characteristics that define collaborative science. And this new scientific paradigm is a key reason why we believe the rate of innovation in the coming decades will eclipse anything that we, or previous generations, have ever experienced. It's also the reason why the New Alexandrians we discuss in this chapter are so central to a robust economic future. To reiterate: The New Alexandrians are individuals, companies, and organizations that recognize the power and importance of openness in today's economy. They are doing more than building a modern equivalent of the world's greatest library. They are building rich collaborative environments and open knowledge infrastructures of all kinds, including open standards, open-content initiatives, open scientific networks, and open research-and-development consortiums.

These are the pillars on which new forms of private enterprises and new twenty-first-century industries will be built. Just as they are the foundation on which a society rich with art, culture, and ideas will flourish.

THE SHARING OF SCIENCE

Call it collaborative science, or even Science 2.0. The Enlightenment accomplished real alchemy, turning research into knowledge by spawning the practice of open scientific publishing. But a centuries-long trend toward openness did not stop there. Today a new scientific paradigm of comparable significance is on the verge of ignition, inspired by the same technological forces that are turning the Web into a massive collaborative work space.

Just as collaborative tools and applications are reshaping enterprises, the new Web will forever change the way scientists publish, manage data, and collaborate across institutional boundaries. The walls dividing institutions will crumble, and open scientific networks will emerge in their place. All of the world's scientific data and research will at last be available to every single researcher—gratis—without prejudice or burden.

Unrealistic you say? Not really, when you consider that conventional scientific publishing is both slow and expensive for users, and that these issues, in turn, are increasingly big problems in science. Visit any campus today and you'll hear ever louder vocal cries for the old paradigm to be swept aside. As new forms of peer collaboration and open-access publishing emerge, this looks more likely by the day. Before we describe this new paradigm, however, let's briefly review the problems.

Traditional journals aggregate academic papers by subject and deploy highly structured systems for evaluating and storing the accumulated knowledge of a scientific community. Each paper is peer reviewed by two or more experts, and can go through numerous revisions before it is accepted for publishing. Frustrated authors can find their cutting-edge discoveries less cutting edge after a lumbering review process has delayed final publication by up to a year, and in some cases longer. With the pace of science increasing today, that's just not fast enough.

The other problem is that the vast majority of published research today is only available to paid subscribers. Ever increasing subscription fees, meanwhile, have made this research less accessible. What's worse is that these impediments to access persist despite the availability of much cheaper electronic publishing methods. Though an unlimited number of additional readers could access digital copies of research at virtually no additional cost, publishers hold back for fear of creating a Napster-like phenomenon.

No doubt these problems are hangovers from a world of physical distribution and a much more limited volume of publishing. The current publishing regime emerged in seventeenth-century Europe, when the pace of discovery was glacial by twenty-first-century standards. Scientific journals provided the primary infrastructure for scholarly communication and collaboration. Apart from annual academic symposiums, journals were *the* place where scientists could find out about, engage with, and carefully critique each other's work. Publishing journals was expensive, entailing significant capital and operational costs.

As the scientific endeavor swells in scale and speed, however, a growing number of participants in the scientific ecosystem are questioning whether the antiquated journal system is adequate to satisfy their needs. New communication technologies render paper-based publishing obsolete. The traditional peer-reviewed journal system is already being augmented, if not superseded, by increasing amounts of peer-to-peer collaboration.

Science Goes Large Scale

Organizing the pursuit of knowledge in a peer-to-peer fashion is certainly nothing new in science. But recent research suggests that collaboration is exploding. One study conducted by the Santa Fe Institute found that the average high-energy physicist now has around 173 collaborators. The same study found that the average number of authors per scientific paper has doubled and tripled in a number of fields. A growing number of papers have between two hundred and five hundred authors, and the highest-ranking paper in the study had an astonishing 1,681 authors.[7]

Knowledge aggregators need to accommodate new realities, such as the growing use of massive online databanks and the rise of large-scale Internet-mediated collaborations. Take the Large Hadron Collider (LHC) experiment at the European Council for Nuclear Research (CERN) for example. Starting in 2007, the world's largest particle accelerator is expected to begin producing petabytes of raw data each year—data that will be preprocessed, stored, and analyzed by teams of thousands of physicists around the world (note: a petabyte is one quadrillion bytes—in other words, quite a lot of data!). In this process, even more data will be produced. There will be a need to manage hundreds of millions of files, and storage at hundreds of institutions will be involved.

Then there is the Earth System Grid (ESG), an experimental data grid that integrates supercomputing power with large-scale data-and-analysis servers for scientists collaborating on climate studies. Once the first of its kind, the project is building a virtual collaborative environment that links distributed centers, users, models, and data throughout the United States. Data for the project is being collected from a wide range of sources, including ground- and satellite-based sensors, computer-generated simulations, and thousands of independent scientists uploading their files. Specialized software applications run on the grid will accelerate the execution of climate models a hundredfold and allow scientists to perform high-resolution, long-duration simulations that harness the community's distributed data systems. The ESG's founders anticipate the project will revolutionize our understanding of global climate change.

Projects like these have inspired researchers in many fields to emulate the changes that are already sweeping disciplines such as bioinformatics and high-energy physics. Take astronomy. The editors of *Nature* recently observed, "A decade ago, astronomy was still largely about groups keeping observational data proprietary and publishing individual results. Now it is organized around large data sets, with data being shared, coded and made accessible to the whole community."[8]

As large-scale scientific collaborations become the norm, scientists will rely increasingly on distributed methods of collecting data, verifying discoveries, and testing hypotheses not only to speed things up, but to improve the veracity of scientific knowledge itself. Rapid, iterative, and open-access publishing will engage a much greater proportion of the scientific community in the peer-review process. Results will be vetted by hundreds of participants on the fly, not by a handful of anonymous referees, up to a year later. This, in turn, will allow new knowledge to flow more quickly into practical uses and enterprises.

In fast-moving disciplines like high-energy physics and bioinformatics, this collaborative way of aggregating and reviewing work is already becoming a reality. In 1991, Paul Ginsparg established arXiv as a public server where physicists could post digital copies of their manuscripts prior to publication. While beginning life as a vehicle for sharing preprints in theoretical physics, it quickly became the principal library for a large fraction of research literature in physics, computer sciences, astronomy, and many mathematical specialties.

"I was originally anticipating about one hundred submissions per year from the roughly two hundred people in the one little subfield it originally covered," explains Ginsparg. "But there were multiple submissions per day from day one, and by the end of the year a few thousand people were involved."[9] Today more than half of all research articles in physics are posted here. And they keep on coming in at a rate of about 4,500 new papers every month. Users can even get RSS feeds that alert them when new research is published in their field.

Dr. Paul Camp of Spelman College, an avid user of the site, says that "[arXiv] is way faster than the traditional publication cycle." Yet the self-organizing community emerging around arXiv manages to preserve the elements of peer review that matter. "What we want is valid, peer-reviewed information," says Camp. "What does it matter if that occurred by means of an editor farming an article out for review, or by direct feedback from the community of people interested in a topic by e-mail in response to your preprint on arXiv? It amounts to the same thing."[10]

Recent efforts such as Google Book Search, the Public Library of Science, and the World Digital Library are now building on the open-access concept. These projects are aggregating vast repositories of scientific research and human culture in easily accessible forms. New science results that might have been available only to deep-pocketed subscribers will now be widely and freely available for education and research. Older resources that might otherwise have wallowed in dusty archives will be given new life and new audiences in digitized formats.

When fully assembled, open-access libraries will provide unparalleled access to humanity's stock of knowledge. Improved access to knowledge, in turn, will help deepen and broaden the progress of science, giving everyone from high school students to entrepreneurs the opportunity to tap its insights.

Collaborative Science in Action

Digital libraries are only the first step in modernizing scientific research and publishing. More profound breakthroughs will come as scientists come to rely less on the "paper" as the prime vehicle for scientific communication and more on tools such as blogs, wikis, and Web-enabled databanks. Blogs such as Bioethics, CancerDynamics, NodalPoint, Pharyngula, and RealClimate suggest that at least a handful of scientists, especially of the younger generation, are already embracing new forms of communication.

Scientists involved in OpenWetWare, an MIT project designed to share expertise, information, and ideas in biology, are heralding the arrival of Science 2.0. Twenty labs at different institutions around the world already use the wiki-based site to swap data, standardize research protocols, and even share materials and equipment. Researchers speculate that the site could provide a hub for experimenting with more dynamic ways to publish and evaluate scientific work. Labs plan to generate RSS feeds that stream results as they happen, and use wikis to collaboratively author/modify reports. Others have suggested adopting an Amazon-style reader review function that would make the peer review process quicker and more transparent.

Meanwhile, over at the European Bioinformatics Institute, scientists are using Web services to revolutionize the way they extract and interpret data from different sources, and to create entirely new data services. Imagine, for example, you wanted to find out everything there is to know about a species, from its taxonomy and genetic sequence to its geographical distribution. Now imagine you had the power to weave together all the latest data on that species from all of the world's biological databases with just one click. It's not far-fetched. That power is here, today.

In a recent editorial on scientific data issues, the editors of *Nature* (one of the world's leading scientific publications) suggest that to harness the power of Web services, scientific institutions will need to rethink the way they collect and manage data.[11] Web services only work if computers can get real-time access to data. Many large public databases like GenBank already allow unimpeded access to their data. But, *Nature* claims, many research organizations still cling to outmoded, manual, data permission policies, and this thwarts the development of Web services.

Scientists invest heavily in collecting data, so it's understandable that many feel justified in retaining privileged access to it, says *Nature*. But there are also huge amounts of data that do not need to be kept behind walls. And few organizations seem to be aware that by making their data available under a Creative Commons license, they can stipulate both rights and credits for the reuse of data, while allowing uninterrupted access by machines.

As Web services empower researchers, *Nature*'s editors rightly point out, the biggest obstacle to fulfilling such visions will be cultural.[12] "Scientific competitiveness will always be with us," they say, "but developing meaningful credit for those who share their data is essential to encourage the diversity of means by which researchers can now contribute to the global academy."[13]

These problems are transitory. Over time cultural inertia will give way to new and improved ways of working and collaborating. Institutional silos, nearsighted data policies, and the static, labor-intensive undertaking of crafting scientific papers will come to represent jumbo-size stumbling blocks in the path of networked scientific communities that thrive on open and rapid communication. Like a river torrent that washes away debris, the flood of peer-to-peer networking in scientific communities will dispose of obsolete policies and practices.

Large open collaborations like the Human Genome Project, to be sure, would not have been possible in today's time frames without the Internet and the emergence of increasingly distributed systems for aggregating, reviewing, and disseminating knowledge. True, there will always be aspects of scientific inquiry that are painstakingly slow and methodical. But as the pace of science quickens there will be less value in stashing new scientific ideas, methods, and results in subscription-only journals, and more value in wide-open collaborative knowledge platforms that are refreshed with each new discovery.

THE PRECOMPETITIVE KNOWLEDGE COMMONS

Speaking of the Human Genome Project, it is certainly among the most important scientific endeavors of our time. When efforts to map the human genome began back in 1986, scientists had barely an inkling of how this fundamental part of our existence works—and to a large degree they still don't! But thanks to massive, distributed collaborations across institutions, countries, and disciplines that took over fifteen years to complete, we are now much, much further ahead than we were in 1986.

One thing that scientists have long suspected is that our genes determine things like what we look like, our intelligence, how well we fight infection, and even how we behave. But armed with a fully sequenced genome, scientists are now convinced that these microscopic spirals of DNA amount to something like an operating system for humans. Learning how to "program" this operating system could hold the key to eliminating dreadful diseases such as Alzheimer's, diabetes, and cancer. Applications of this research in fields such as agriculture and ecology could help us end world hunger and take better care of the planet.

But for us the Human Genome Project is important for an additional reason: It helps illustrates our key thesis in this chapter. The Human Genome Project represents a watershed moment, when a number of pharmaceutical firms abandoned their proprietary human genome projects to back open collaborations. By sharing basic science and collaborating across institutional boundaries, these brave companies challenged a deeply held notion that their early stage R&D activities are best pursued individually and within the confines of their secretive laboratories. As a result they were able to cut costs, accelerate innovation, create more wealth for shareholders, and ultimately help society reap the benefits of genomic research more quickly.

So what exactly were these firms up to? We call it a "precompetitive knowledge commons," and we agree it's a bit of a mouthful, but we're talking about something big—a new, collaborative approach to research and development where like-minded companies (and sometimes competitors) create common pools of industry knowledge and processes upon which new innovations and industries build.

Prospecting the Genome

Thanks to these efforts the race to sequence the human genome bequeaths an impressive legacy. GenBank, the National Institute for Health's repository of gene sequences and other related information, is now the world's largest public database of genetic information. It is the culmination of myriad public and private efforts that placed genetic information in the public domain.

This public resource promises to be enormously valuable. It provides an infrastructure of freely available scientific information for millions of biomedical researchers and will spur follow-on innovation for decades. Recent GenBank statistics already demonstrate its growing value. As of August 2009, researchers have collected and disseminated over 100 gigabases. That's 100,000,000,000 "letters" of genetic code from over 165,000 organisms. For a frame of reference, 100 gigabases means 100 billion base pairs of DNA, which is just slightly less than the number of stars in the Milky Way.

Impressive growth and usage statistics, in turn, lend credibility to those who argue that a robust scientific commons is the best way to ensure we realize the full potential of the genomics revolution. David Lipman, director of the National Center for Biotechnology Information, speculates that this thriving knowledge commons will soon give researchers the ability to map and understand the genetic makeup of entire ecosystems, not just the human genome.[14]

Truth is, however, the efforts to sequence the genome could easily have gone the other way. In the wake of controversial court rulings that have allowed patent rights over genetic information since the early 1980s, for-profit and nonprofit entities became enthusiastic participants in the patent system. By the mid to late 1990s researchers and industry participants feared that patents on large amounts of DNA sequence data would confer potentially very broad rights to exclude others from working on scientific and therapeutic applications. As tens of thousands of patent applications flooded the United States and European Union patent offices, debate erupted over the patentability of isolated gene fragments and the future of biomedical research.

Biomedical researchers feared (and still do) that access issues would erode the culture of open science and impede scientific progress. Some 20 percent of the human genome was already under private ownership, including the genes for hepatitis C and diabetes. The owners of these patents now influence who does research and how much it will cost them, playing a disproportionate role in determining the overall rate and direction of research in these areas.

While scientists worry about academic freedom, pharmaceutical firms worry about paying excessive licensing fees to a new class of competitors—the biotechnology companies—that have emerged at the interface between academic and commercial research. By the late 1990s only a handful of firms had mastered the technologies to synthesize, analyze, and annotate the escalating volumes of data produced by public and private gene-sequencing projects. Big Pharma was eager to mine this information for potential blockbusters but lacked the requisite capabilities. With few suppliers and everyone moving quickly in the race to prospect the genome, biotech firms could command premium prices for the latest information and tools. Many firms used this leverage to negotiate "reach-through rights" that allowed them to lay claim to future discoveries.

Both the academic and commercial communities warned that locking up significant portions of molecular biology was raising costs and lowering the efficiency of drug discovery. As patents proliferated, R&D budgets were rising to inefficient levels, and biotechnology companies, pharmaceutical firms, universities, government entities, purchasers of health care, and the legal system were getting entangled in expensive and damaging struggles for the associated economic benefits.

In short, the industry was in crisis, and there seemed to be little that any one player could do about it except join in the genomic gold rush.

Big Pharma Fights Back

One company, however, saw another option that could rewrite the rules completely. In 1995, Merck Pharmaceuticals and the Gene Sequencing Center at the Washington University School of Medicine announced the creation of the Merck Gene Index, a public database of gene sequences. Merck immediately released 15,000 human gene sequences into the public domain and announced that it would characterize and make freely available as many gene sequences as possible. Under the terms of the agreement, no one could gain advance access to—or even delay or restrict the release of—any of the sequence data from Merck and Washington University. This included Merck researchers, who gained access to the data via the same public databases available to all interested researchers.

By 1998, Merck and Washington University had published over eight hundred thousand gene sequences. As long as the gene sequences were public no company could lay claim to them. The strategy appears to have worked: Recent evaluations of the threat of gene sequence patents to biomedical research progress suggest that the gene index (along with other parallel public efforts) has significantly eased the gold rush dynamic. But why would Merck make this investment, which, according to one estimate, cost them several million dollars?

Dr. Alan Williamson, former vice president of research strategy with Merck, explains it in philanthropic terms: "Merck's approach is the most efficient way to encourage progress in genomics research and its commercial applications. By giving all research workers unrestricted access to the resources of the Merck Gene Index, the probability of discovery will increase. The basic knowledge we and others gain will lead ultimately to new therapeutics for a wide range of disease, while providing opportunities—and preserving incentives—for investment in future gene-based product development."[15]

Nice sentiments, but a subtle element of competitive sabotage underlies this apparently soft strategy. Like many pharmaceutical firms, Merck sees gene sequences as inputs rather than end products. Their business is developing and marketing drugs, not hawking genetic data and research tools. By placing gene sequences in the public domain, Merck preempted the ability of biotech firms to encumber one of its key inputs with licensing fees and transaction costs.

Fortunately for Merck, other pharmaceutical firms shared its concern over patents on upstream genetic information. Similar collaborative projects that built on Merck's approach were soon launched on a much larger scale. In 1999, the SNP Consortium was established as a collaboration of eleven pharmaceutical companies, a nonprofit institution, and two IT firms. This unique joint venture brought highly competitive companies together—which rarely share any information, let alone information from a potentially path-breaking basic science initiative—to produce what the founders call "a public biological blueprint for all human life." Their common goal: to hasten a new era of "personalized medicine" in which treatment is tailored to an individual's unique genetic profile.

Many pharmaceutical executives believe that thanks to advances in gene technology, the key to future blockbuster therapies is identifying which drugs work best for which patients. Scientists are increasingly convinced that minute genetic differences largely account for people's different health traits and explain why a drug works for one person but has no effect—or ill effects—on another.

In the mid-1990s, scientists discovered that tiny chemical landmarks inside or near genes are posted at regular intervals along the DNA molecule, like road signs and mile markers on a stretch of highway. These landmarks, called single nucleotide polymorphisms (SNP), could be used to create a catalog of the ever so slight genetic variations that make some individuals susceptible to disease. As Francis Collins, a director at the National Human Genome Research Institute, put it, "SNPs serve as a blinking light on DNA sequences showing there is something very interesting here—for example, something that is contributing to diabetes."[16]

The SNP Consortium set out to identify the hundreds of thousands of chemical landmarks along human DNA. Alan Williamson, then recently retired from Merck, helped organize the initial talks among the consortium's partners. He recalls the excitement: "Suddenly, there was going to be a genetic map powerful enough to define which patients respond to a given drug versus those which don't respond to a given drug. . . . It would allow doctors to tailor treatments to patients more exactly than ever before."[17]

The initial goal was to map 300,000 common SNPs. At the completion of the project in 2001, 1.8 million had been mapped. To achieve this goal, the consortium invested approximately $50 million to pay university researchers to discover SNPs and place them in the public databanks. The consortium also filed patents to establish priority and obtain legal standing to contest other filings. Applications were abandoned once the SNPs were securely in the public domain.

Now that the SNPs have been mapped, the harder interpretive discovery work leading to new diagnostics and therapies is beginning. As a testament to its effectiveness, a generous flow of follow-on innovation is proceeding in the wake of the SNP project.

Commercial and academic scientists are currently using the map to filter rapidly through genetic profiles from thousands of patients to uncover which of the one hundred thousand or so genes that make up human DNA predispose people to such common but hard to treat ills as diabetes, depression, cancer, arthritis, Alzheimer's, and heart disease. The underlying biological causes of these illnesses remain largely mysterious, but if uncovered, this knowledge could lead to a treasure trove of new treatments.

The Value of Collaborative Discovery

But why collaborate when competition would let the winner extract proprietary gains? And why put this valuable information in the public domain? Why not limit disclosure to the consortium's members? As with Merck's Gene Index, there is blocking value in making public valuable but noncore information. The consortium's initiative competed directly with the biotech companies (including Incyte, Millennium Pharmaceuticals, and France's Genset) that were making their own proprietary catalogs of genetic landmarks. Though wary of sharing their valuable data with rivals, the consortium's members worried even more about biotech companies' projects. Daniel Cohen, former lead scientist at Genset, claimed at the time that Genset's plan to patent SNPs and sell them to the highest bidder would net between $50 million and $100 million a patent.

SNP members deny any concerted attempt to disrupt biotech competitors. "The idea here isn't to restrict the ability of biotech firms or anyone else to patent genes," says Williamson. "The idea is to make sure the underlying map we all need to find genes is available to anyone who wants to use it."[18]

But it's in the interests of Big Pharma to level the playing field for gene-hunting biotech firms, large drug companies, and academic scientists. The competencies of the consortium's members lie overwhelmingly in drug development, approval, and marketing. They are collectively better off competing to bring valuable end products to market than competing with the biotech firms in upstream research. Lawyers also reportedly advised consortium members that making the map public would help the companies avoid antitrust problems. In the end, however, the consortium members' big prize for collaborating is not the blocking value, but the benefits of speeding the industry toward personalized medicine. Before agreeing to collaborate, many consortium members were already building their own proprietary SNP maps. Under the leadership of Alan Williamson, they realized that a common map was crucial to the success of personalized medicine.

As Allen D. Roses, senior vice president of pharmacogenetics for GlaxoSmithKline, explains, "It was crucial that we had something whose accuracy we all agreed upon. If each of us had produced our own map, it would, for one thing, have taken much longer to create, and it would have been very unlikely that the companies would have accepted one another's map as being valid."[19] Among other things, the Food and Drug Administration (FDA) also needed to know that the map was accurate, reliable, and accepted by the scientific community.

By fusing corporate resources with the relatively low-cost contributions of academic scientists—which after all could only be bought for a low price if the data remained public—the consortium was able to discover many more SNPs than it imagined: 1.5 million more! And they did so in a fraction of the time it would have taken a single firm. This meant resources that may have been wasted pursuing duplicate research could be redirected toward other goals, namely the pursuit of follow-on diagnostics and therapeutics.

OPEN SOURCE DRUG DISCOVERY

Despite the major scientific achievements of sequencing of the human genome, progress in other areas of biomedical research and drug development has so far been disappointing. No new broad-spectrum antibiotics have been marketed in almost forty years, and many forms of cancer, as well as chronic diseases and disorders such as Alzheimer's, Parkinson's, and schizophrenia still lack effective and well-tolerated treatments. There has been almost no research on tropical diseases such as malaria and typhoid, the burden of which falls almost entirely on the world's poorest populations. In fact, only 1 percent of newly developed drugs will help the millions of people in Africa who die annually from these diseases.

Even the blockbuster drug business is suffering. In 2002, the FDA approved only seventeen new molecular entities (NME) for sale in the United States—the lowest since 1983, and a fraction of the fifteen-year high of fifty-six NMEs approved in 1996. In 2003, the FDA approved twenty-one NMEs, of which only nine were designated as "significant improvements" over existing drugs. This decline occurred despite a substantial increase in R&D spending: Between 1995 and 2002 U.S.-based pharmaceutical companies roughly doubled their R&D expenditures, to about $32 billion.[20] Numbers like these have the popular press and trade journals talking about "dry," "weak," or "strangled" pipelines, and a productivity crisis with dire consequences for investors (who can expect "permanently lower multiples"), taxpayers, patients, and insurers, who will have to pay an ever higher bill to maintain the pace of technological progress in the industry.

Dr. Frank Douglas, former executive vice president and chief scientific officer of Aventis, agrees that there are many concerns to address: "The productivity of large pharmaceutical innovation has decreased," says Douglas. "We lack the ability to properly predict the side effects of new compounds, and we don't have good ways to monitor and assess them once they are in the market. Pricing models have become untenable. So has the 'blockbuster' mentality. Across the board, a lot of old models really need to be examined."[21]

Indeed, as increased research spending collides with pressure to contain health care costs—and as alarm grows at the seemingly callous neglect for diseases that disproportionately affect the world's poor—the factors that affect the efficiency of drug discovery and development have rightfully come under scrutiny. The promise of biomedical research to relieve human suffering and create wealth has never been higher. But the ability of the industry to deliver on this promise depends critically on its ability to control costs, marshal resources effectively, and manage its knowledge base efficiently.

The Open Source Opportunity

Having witnessed what Linux has done for software production it seems natural to wonder whether a flurry of open source activity could unleash a similar revolution in the life sciences. What if the drug discovery process, for example, was opened up so that anyone could participate, modify the output, or improve it, provided they agree to share their modifications under the same terms? Could the collective intelligence of the life sciences community be harnessed to enable a more coordinated and comprehensive attack on the intractable diseases that have so far stymied the industry? Could opening up the process to tens of thousands of volunteer researchers lower the cost of drug development to the point where the resulting medicines are within reach of the world's poor? A small number of visionaries think there is an enormous opportunity here. But no one is suggesting it will be easy.

For one, there are fundamental differences between creating software and developing new drugs. Software production is easy to break up into bit-size pieces that can be carried out on a laptop while sitting in Starbucks. Drug development is harder to parse out and requires access to expensive laboratory instruments. Software projects can be completed in months, or even days and weeks. A typical drug currently takes ten to fifteen years and an average of $800 million to develop. Making software inventions commercially viable is easy and inexpensive—just post it on the Internet. Biological inventions take years of painstaking clinical trials and a healthy dose of regulatory know-how to reach that point. All these factors make drug development less hospitable to peer production than software.

On the other hand, there is much that unites open source programmers and the biomedical research community. Both communities share similar goals (free software and accessible medicines) and are driven by similar motivations (such as reputation and learning). They share strong community ethics, such as reciprocal sharing and collaborative discovery. And most of the people who contribute to collaborative projects in software and bio-medicine are either paid to do so directly (i.e., as employees of companies and universities), or do so in their spare time while earning a living in some facet of the industry.

The fact that drug discovery is increasingly conducted in computer networks rather than in test tubes opens another window to open source activity. Indeed, many of the tools for sifting through the mountains of genomic data produced by the Human Genome Project are already available as open source. Bioinformatics.org, one of several hubs for collaboration in the biomedical community, hosts over 250 active projects that extend open source software development practices to the biological research databases and software tools. Freely available genomic search-and-comparison algorithms such as BLAST (Basic Local Alignment Search Tool) are becoming de facto standards in the community.

These factors suggest that peer production will have a significant role to play in drug discovery, particularly in the early stages, where the minds of thousands of scientists can be harnessed to identify promising candidates. But the costs and risks of drug development escalate as promising drug candidates move farther down the pipeline. The deep investments firms make at these stages are premised upon the availability of patent protection, which provides a period of exclusivity in the marketplace. The need to obtain patent protection, in turn, drives firms to throw up iron curtains around their research the moment they get close to a viable drug candidate.

Today a variety of nonprofit initiatives are seeking answers to these conundrums. Public-private partnership models that pool the resources of Big Pharma, philanthropists, government, and nongovernmental organizations currently offer the most hope for neglected diseases. Though a variety of different partnership models are plausible, the most promising would link upstream open source drug discovery efforts to downstream consortiums that usher good candidates through the later stages of development. This way companies would minimize R&D costs by involving partners at various stages of the development process, particularly at the costly clinical stage where suitable partners in the public sector can take over.

So far, projects led by the Institute for OneWorld Health, the Gates Foundation, and the Drugs for Neglected Diseases Initiative (among others) are making significant headway on diseases such as malaria and tuberculosis. Companies such as GlaxoSmithKline, Novartis, AstraZeneca, and Sanofi-Aventis have recently become enthusiastic participants in these initiatives. They may not stand to make any profits, but they can at least enhance their corporate images while taking advantage of a low-risk, low-cost route to getting established in developing-country markets. What's more, if open source drug discovery works, then these companies can apply a similar formula to cut costs and increase innovation in their ailing block-buster businesses.

RETHINKING INDUSTRY-UNIVERSITY PARTNERSHIPS

Innovation can come from many sources and in many different forms. Smart companies realize that remaining competitive means innovating in all aspects of their business. Innovation, after all, is not just a product of science and invention. Cocreating with customers, peer producing value with partners, and optimizing supply chains (among other things) are equally pivotal.

Still, advancing the basic sciences is really the only way to guarantee that industries will continue to be innovative over the long term. Imagine farming without organic chemistry, or medicine without microbiology, or electronics, computing, and semiconductors without quantum mechanics. Without new insights and advances in the underlying disciplines our stock of knowledge becomes stale. If the well of knowledge dries up, so too does innovation.

Until recently firms took on a large share of the responsibility for advancing the underlying sciences. But, as explained in Chapter 4, they engaged in too much invention for invention's sake, while their R&D proceeded at a leisurely, academic-like pace.

Some of this basic research yielded large dividends for society and their shareholders: Think of DuPont's investments in basic chemistry that led to the invention of synthetic rubber, or the invention of the transistor in AT&T's Bell Labs. But much of it did not translate into immediate opportunities to market new products and services. Lack of a clear return on investment led to a dramatic scaling back of basic science in corporate R&D departments beginning in the late 1980s and continuing into the 1990s.

Today it is more important than ever that R&D activities are fast and efficient and earn a clear ROI. Innovators will still need to know the underlying sciences, but their primary aim in-house cannot be to further the science. For that they will increasingly rely on partnerships with universities and other research organizations, while corporate research teams use their skills and resources to move quickly to practical application. Indeed, smart firms see university partnerships as a nimble and cost-effective means for detecting and launching disruptive innovations.

The problem for many mature companies is that the very commercial success of their products increases their dependency on them. Making radical changes in the product's capabilities, underlying architecture, or associated business models could cannibalize sales or lead to costly realignments of strategy and business infrastructure. It's as though popular and widely adopted products become ossified, hardened by the inherent incentive to build on their own successes. The result is that entrenched industry players are generally not motivated to develop or deploy disruptive technologies, as Harvard Business School professor Clayton Christensen has pointed out.

So success breeds complacency. R&D departments have often stopped learning about alternative technologies and channeled their resources into refining components, adding new features, or tweaking the product architecture. This strategy of marching down a well-defined product road map may pay dividends for some time. But complacency creates two kinds of vulnerabilities.

The first is that research conducted in pursuit of well-defined product road maps rarely leads to entirely new business lines or significant changes in corporate strategy. Yet both new business lines and periodic changes in corporate strategy are required to keep employees refreshed and to support long-term company growth.

The second is that focusing narrowly on improving existing products will inevitably lead firms to fail to detect disruptive innovations that may threaten the product road map itself. Ideally, companies will detect such innovations long before they reach the marketplace, giving themselves enough time to turn potentially fatal developments into distinct competitive advantages.

The problem is that the kind of exploratory research required to renew corporate strategies and detect disruptive innovations is also the most costly and risky. David Tennenhouse, a renowned technologist and former vice president of Intel's corporate technology group, thinks these are costs and risks that are best shared through a new open and collaborative model of industry-university partnerships.[22]

Intel's Open University Network

Having spent much of his career working at DARPA (Defense Department Advanced Research Project Agency), structuring close collaborations among public and private agencies is something that Tennenhouse knows a lot about. He took that knowledge to Intel, where he built a highly successful approach to managing Intel's university partnerships.

Tennenhouse identifies a couple of reasons for the growing relevance of university research. "The number of talented electronics and IT researchers has grown substantially, and the talent pool is widely dispersed," says Tennenhouse. "Ideas now flow through leading universities and their faculty rather than any one industrial lab, however prestigious."[23]

Accelerating technological change and heightened competition from Asian semiconductor firms are also putting the heat on Intel. Close cooperation with leading universities helps Intel maintain its edge, while spreading the upfront costs of R&D across a much broader research ecosystem. Tennenhouse says that by leveraging its university connections skillfully, it gains access to the results produced by the bulk of the research community.

In the spring of 2001, for example, Intel established exploratory research labs adjacent to the University of California at Berkeley and the University of Washington in Seattle. Two more labs, near Carnegie Mellon University and the University of Cambridge (UK), were added later. Intel selected leaders in the research areas it wanted to explore who had a strong track record of collaborating with industry, and whose faculties collaborated well with one another.

Each lab houses twenty Intel employees and twenty university researchers. "Company and university researchers work side by side," says Tennenhouse, "and communicate their findings instantaneously rather than waiting to present them first via formal channels, such as conferences and publications." Each lab has a unique research focus—from ubiquitous computing to distributed storage.

When a promising research thread is detected, Intel puts a coordinated set of efforts in motion that includes additional grants to leading university researchers and the initiation of its own complementary projects. At the same time, Intel works closely with its corporate venture group to identify and invest in promising start-ups in each new sector.

The key to the program's success in transferring people and technology across university/industry boundaries is funding multiple projects at a time. Intel sets things up so that university and Intel research teams work in parallel, while meeting regularly to exchange results. This way, says Tennenhouse, "the researchers at different institutions compete among themselves but also work together to achieve the program objectives. These cycles of competition and hybridization," he adds, "lead researchers to quickly adopt the best of each other's ideas."

In the four years since the first exploratory lab was launched, Intel Research has matured more quickly than expected, accelerating research in a number of key areas. Already, five strategic research projects, in the areas of polymer storage, microelectromechanical systems (MEMS), optical switching, inexpensive radio frequency, and mesh networking have been transferred downstream toward product development. "The labs are generating strong intellectual results (as evidenced by publications at premier conferences), our efforts in the areas of sensor networks and PlanetLab have been widely acknowledged, and our ubiquitous computing team is recognized as one of the best," says Tennenhouse.

Making the Most of University Partnerships

To replicate some of Intel's successes and make the most of university partnerships in your firm, we recommend adopting the following principles.

Use industry-university partnerships to shake up product road maps
Though incremental movement is a powerful and important feature of innovation, focusing solely on incremental product improvements can easily lead to stagnation.[24] With a few notable exceptions, corporate research teams have been unable to sustain a high level of success over time. Beyond a decade, their agendas, once bold and innovative, become conservative and incremental. To fight attrition, Intel uses industry-university partnerships to deliberately introduce disruptive elements into its strategy.

Make sure the collaboration is a win-win

Cash-constrained university departments generally welcome industry sponsorship of their research programs. But such partnerships are not free from controversy, so it's wise to take note of the following considerations. First, don't poach all of the university's top staff. Universities live and die by the quality of their faculty—star faculty attracts students and funding and generates top rankings. Second, be sensitive to the need for faculty to publish and conduct ongoing research. Published research is the key metric by which faculty are evaluated, both by their employers and the wider research community. Finally, create lasting relationships between company and university researchers that continue to create value long after the formal partnership itself has ended. Intel researchers, for example, often keep in touch with faculty members, and occasionally lean on them when they're running into difficulties.

Deepen and broaden collaboration across research communities

Many industry-university partnerships are structured such that individual project teams at different institutions work in isolation. Yet Intel has found that some of its most promising insights and applications may flow from unexpected synergies that arise when teams get together to discuss their research. With sufficient critical mass and geographical reach, collaboration across institutions could even jump-start whole new research communities.[25] Tennenhouse describes this practice as "reverse technology transfer." Instead of transferring technologies from the university to Intel's business units, Intel sometimes reverses the flow, moving technology back upstream to university researchers. Doing so allows Intel to foster vast research communities with the scale to collectively address huge research challenges of strategic importance to the company.

Keep the science open and the applications proprietary

Rather than wrangle over who gets to control and exploit the fruits of joint research efforts, Intel and its academic partners sign on to Intel's open collaborative research agreement, which grants nonexclusive IP rights to all parties.[26] This way both sides retain their freedom to engage in further research, develop new products, and partner with other players. This may sound like martyrdom in the name of openness, but in fact the benefits of casting a wide net for new ideas and learning rapidly from the external research ecosystem greatly outweigh the advantages gained from keeping the research proprietary. "Proprietary advantage," says Tennenhouse, "is more effectively obtained in the downstream stages of a project, as the work moves toward technology and product development."

Learn from "proxy" customers—early and often
One of the elements so often missing in exploratory research is the customer's perspective. Intel strongly encourages each of its project teams to place interim results and prototypes into the hands of proxy customers as soon as it is practical to do so. These early users provide feedback about which aspects of their research are most valued by customers (or not) and which applications are worth pursuing—applications that are frequently different from those originally envisioned.

LAYING THE PUBLIC FOUNDATION

Competition through free enterprise and open markets are at the heart of a dynamic economy, but if there is one additional lesson to take away from this chapter it's that we can't rely on competition and short-term self-interest alone to promote innovation and economic well-being. Vibrant markets rest on robust common foundations: a shared infrastructure of rules, institutions, knowledge, standards, and technologies provided by a mix of public and private sector organizations.

A growing number of leaders in the private sector now appreciate the value of a strong public foundation. These New Alexandrians understand that creating a shared foundation of knowledge on which large and diverse communities of collaborators can build is a great way to enhance innovation and corporate success.

Some companies use cross-licensing and patent pools to lower transaction costs and remove friction in their business relationships. Some industries embrace open standards to enhance interoperability and encourage collaboration. Others invest in a precompetitive knowledge commons to boost the productivity of downstream product development. Still others prefer to help weave networks of university partners that will unleash a fertile stream of ideas and inventions that can blossom into new businesses. Regardless of which method—or combination of methods—firms choose, the result is usually the same: a more dynamic and prosperous ecosystem.

Despite this promising flurry of open activity there are still far too many companies, and their allies in public office, that take the public elements of the innovation equation for granted. They see calls to further open up infrastructures for communication and collaboration, to enlarge the public domain, or to create a more balanced intellectual property system as inimical to economic prosperity. Somehow the dismal record of economic development in many developing nations, where such public institutions are weak, has not convinced them otherwise.

These are book-length topics, and it's hard to do them justice here. But the way we manage intellectual property, in particular, affects everything we have discussed in this chapter—and indeed most of the new business models we cover in this book—so it's worth reflecting on the topic. Of course, as authors and business people we recognize that rewarding creativity and investment is central to promoting innovation. In theory, intellectual property law exists to do just that. But expansion in the law's breadth, scope, and term over the last thirty years has resulted in an intellectual property regime that is radically out of line with modern technological, economic, and social realities. This threatens the chain of creativity and innovation on which we (and future generations) depend.

In today's economy we need an intellectual property system that rewards invention and encourages openness—one that fuels private enterprise and sustains the public domain. The existing intellectual property system isn't working as well as it could.

Increasingly vocal critics argue that our knowledge economy has become overprivatized. Scholars such as James Boyle and Lawrence Lessig point out that in recent decades intellectual property rights have been consistently strengthened, while the public domain has become dangerously constricted. These voices need to be heard.

Since the Bayh-Dole Act extended patent eligibility to public research organizations in 1980, for example, property rights have been migrating farther upstream into the realm of basic science. On the one hand, property rights in basic research offer the promise of substantial economic gain from increased commercialization of inventions. On the other hand, commercialization could erode the culture of open science that has fueled centuries of scientific discovery.

Science and commerce depend upon the ability to observe, learn from, and test the work of others. Without effective access to data, materials, and publications, the scientific enterprise becomes impossible. Recent studies show a disturbing trend: Increasing secrecy, pressures to patent, cumbersome technology-transfer agreements, and complex licensing structures are making it hard for scientists to share research data. In a recent survey by the American Association for the Advancement of Science, 35 percent of academic researchers reported difficulties that had affected their research because they were denied access to data, while 76 percent of scientists working in industry reported similar problems.

Concerns about access are serious. Strong, well-funded academic research institutions are a pillar of any nation's commercial success. In the United States, National Science Foundation figures show that while American academic institutions perform 13 percent of national R&D (spending about $36 billion), they perform 54 percent of all basic research. A significant portion of this basic research (50 percent in 2001) goes into the biological and medical sciences—a key frontier for scientific discovery and economic growth.

By comparison, large, research-driven pharmaceutical firms like Merck with an annual R&D budget of about $3 billion conduct less than 1 percent of the biomedical research in the world. To gain access to the remaining 99 percent of biomedical research, pharmaceutical firms tap into the research conducted in universities and public research organizations around the world. If royalties or restrictive licensing conditions inhibit public researchers' access to patented research tools, then the industry's opportunity to harvest this research diminishes.

The Balancing Point

Finding the right balance between the public foundation and private enterprise is key to the long-term competitiveness of firms and economies. We have to be able to apply existing knowledge to generate new knowledge. At the same time society must elicit the private investment needed to translate new knowledge into economic and technological innovations that contribute to social well-being.

In short, we must encourage innovation without eroding the vitality of the scientific and cultural commons. We need an incentive system that rewards inventors and knowledge producers *and* encourages dissemination of their output.

The hard questions are as follows: How much protection is enough or too much? What's the right balance between private enterprise and the public domain? And what will best achieve that balance—market mechanisms or government intervention?

Public policy reforms are undoubtedly warranted. In intellectual property law, many practitioners are calling for the courts, Congress, or international treaties to roll back—or at least counterbalance—property rights. Well-targeted legal measures could significantly reduce some of the current costs and uncertainties that have accompanied the recent wave of privatization.

But curbing the excesses of the intellectual property rights system will require a broader portfolio of initiatives that includes collective action by firms and nongovernmental organizations, and above all, a new way of thinking about openness and sharing. Indeed, while policy measures are being debated, smart firms should be taking action.

The prompt publication of data, methods, and source code in the life sciences industry, for example, appears to have been a powerful constraint on patenting. To the extent that this "push back" preserves commercial freedom of action on the one hand, and freedom of inquiry on the other, it contributes to the long-run performance of biomedical research and the pharmaceutical industry. Indeed, if academics were to be squeezed out of a patent-laden field, industry would be cutting off its most important lifeline.

On the other hand, the scorched-earth strategy of placing data, methods, and source code in the public domain may have undesirable consequences. If patents become more difficult to obtain, commercial researchers may become secretive to protect their investments, thereby limiting access to important knowledge and making duplicated research more likely. Even worse, fundamental aspects of the industry's infrastructure may suffer due to chronic underinvestment.

Balancing these concerns is critical to maintaining the health of the life sciences industry ecosystem. Analogous concerns arise in any industry where R&D activities are distributed among upstream and downstream firms, and at some stage, a nonprofit research community—a scenario that describes almost all science-intensive industries today.

This brings us to one last lesson, which pertains to the importance of choice and balance. Companies can't open the kimono all of the time. Companies need to defend their assets and work hard to create proprietary advantages. Pharmaceutical firms may harness openness in the early stages of drug discovery. But nobody is giving up patent rights over end products. Indeed, every member of the SNP Consortium is fighting tooth and nail to be the first to get new drugs to market.

Every firm has to reach its own set of conclusions about the appropriate balancing point. Indeed, it is essential to the competitive and evolutionary process that rival forms of strategy and organization can do battle. There is something truly inspiring about a world where the clash of worldviews between Microsoft and IBM or Big Pharma and the biotech firms can play itself out in the marketplace. This, as IBM strategist Joel Cawley put it, "will generate an evolving set of commons, an evolving set of protected areas, and an evolving set of walled gardens."[27] It is the vitality of this evolution that is important. So long as the playing field remains level, it is one reason to remain optimistic about the future.

7. PLATFORMS FOR PARTICIPATION

All the World Is a Stage, and You're the Star

In May 2005, Paul Rademacher was trying to find a house in Silicon Valley for his job at DreamWorks Animation. He grew weary of the piles of Google maps for each and every house he wanted to see, so he created a new Web site that cleverly combines listings from the online classified-ad service craigslist with Google's mapping service. Choose a city and a price range, and up pops a map with pushpins showing the location and description of each rental. He called his creation housingmaps.

While a useful tool for helping people find a place to live, on the surface it hardly seems groundbreaking. And yet, Paul Rademacher's site quickly became a poster child for what the new Web is becoming, not because of what it was, but for how it was created. Housingmaps was one of the Web's first mashups.

After housingmaps popularized the concept, similar mashups were announced on a daily basis. Web sites dedicated to tracking the mashup phenomenon now catalog nearly one thousand unique implementations. The majority of these are a variant of what Paul Rademacher did: link a data or content source to a mapping application to build a geographic display of location-based information. Google Map mashups, for example, have emerged to do everything from pinpointing the locations of particular crime sites, to outing celebrity homesteads, to enabling fitness buffs to measure their daily running distance. Or, for the price conscious, there's CheapGas, a service that mashes Google Maps and GasBuddy together to identify gas stations with the lowest pump prices.

Intriguing though they may be, Google Maps mashups are merely the tip of the collaboration iceberg. We are entering a world where vast open platforms for participation provide a foundation on which large communities of partners can innovate and create value. Open platforms are different from the prosumer communities discussed in Chapter 6. In prosumer communities a company develops a strategy to cocreate products with its customers. With open platforms a company creates a broader stage upon which various partners can build new businesses or simply add new value to the platform. Like prosumer communities, ideagoras, and other peer-to-peer phenomena described in this book, open platforms are mass collaboration in action—a bold new way to extend the productive capacity of your business without having to infinitely extend your fixed costs.

How do you know a platform for participation when you see one? The truth is, that's up to you. A platform might be a Web service such as Google Maps. Or, like Amazon, it might include an e-commerce system for warehousing, purchasing, and distributing goods. A growing number of companies are leveraging platforms like these to create on-the-fly partnerships with developers who build value-adding applications. We call these developer ecosystems—fluid webs of business partnerships that are created when a company opens its software services and databases via an application programming interface (API).

The best examples are the developer communities that have formed around eBay, Google, and Amazon. External partners can build tools to leverage the database information, invent new kinds of stores or applications, and generally integrate their business processes. For example, 40 percent of the goods on eBay are now uploaded automatically from the inventory systems of third-party stores that use eBay as an alternative sales channel. Amazon enables thousands of software developers to access its product database and payment services to create their own new offerings. "We're all building this thing together," says Jeff Bezos, referring to Amazon's customers, Web sites with Amazon associate stores, and outside software developers.[1]

Platforms for participation can also include products ranging from a video game console to a car—virtually anything that runs software. Think of the car of the future. The car is not just a vehicle for moving around; it's a place for work, learning, and entertainment, with a series of software programs connected to a network. Now imagine a car with a set of open APIs, allowing thousands of programmers and niche businesses to create custom applications for your car.

The message to businesses is clear: Open up your platforms to increase the speed, scope, and success of innovation. Choose not to open up and you risk ceding the game to more nimble platform orchestrators. The question every business leader in every sector should be asking is: How do I make my organization a platform for participation? How, when, and where do I open up my business? And how do I attract an energetic group of people to share the innovation load?

For individuals and small businesses, the creative and entrepreneurial opportunities to build on these modern-day infrastructures have never been greater. By tapping open platforms you can leverage world-class infrastructures for a fraction of the cost of developing them yourself. The Web—indeed the world—is your stage, so get ready to deliver your star performance.

PLATFORMS TO THE RESCUE

Despite the novelty and potential utility of Web services like housingmaps, many people might understandably find the fact that we can now remix data and software into recombinant creations somewhat banal. Before you write off the phenomenon as another inane Silicon Valley obsession, consider this: Less than four months after housingmaps was released, the mashup concept was deployed in an altogether more serious and ultimately tragic situation. This time the power to remix the Web was harnessed not to showcase a software nerd's cleverness, but to reunite distressed families, find housing, and even save lives in the wake of one of America's worst natural disasters, Hurricane Katrina.

For many Americans the episode leaves a dark stain on the nation's history. Hurricane Katrina ripped into the coastlines of Louisiana, Mississippi, and Alabama on Monday, August 29, 2005 causing more human misery and economic damage than any storm on record. The ruthless and indiscriminate wrath of nature's forces, it turned out, was just a prelude to the real misery. Perhaps more damaging was the astonishingly inept government response that left hundreds of thousands of hurricane victims stranded, without money, food, water, or clothing, separated from their loved ones, and many in desperate need of medical attention.

Yet, out of the chaos, and in the face of official ineptitude, came a powerful story of how an ad hoc team of volunteers from across the country came together to concoct an information management solution that far surpassed anything the local, state, and federal response teams had mustered. At the heart of the volunteer effort was a central repository of survivor information called Katrinalist. The impromptu Web site compiled survivor data from all over the Web into a searchable format that made it easy to identify and locate friends and family members. There were no government grants, official mandates, formal command structures, or elaborate communication protocols; just a loose group of committed individuals under effective grassroots leadership harnessing rudimentary Web services technologies to help those in need.

It was dubbed the PeopleFinder project and here's how it happened. Shortly after Katrina struck New Orleans, people began posting notices in online forums and popular destinations across the Web in hopes of connecting with loved ones. Missing persons locator services soon mushroomed. The Red Cross had its Family Linking site. Craigslist, Yahoo, and Google each had their own services. Indeed, every news network, university, nonprofit organization, and ad hoc evacuee group in the country had seemingly developed its own find-your-loved-ones-here Web site. Though the intentions were good, critical data was hopelessly scattered, making it virtually impossible to guarantee that important information would reach the people who needed it.

Then on Friday, September 2, David Geilhufe rallied a handful of tech-savvy volunteers to help make order out of chaos. Geilhufe, who runs a nonprofit social-software outfit, started scouring various databases and online bulletin boards using an automated process called "screen scraping" that captures the relevant information for each person—name, location, age, and description—and collates it in a central database. Geilhufe and his team even authored an open source data spec for organizing the missing persons information called the PeopleFinder Interchange Format.

Still, there was only so much that the automated screen-scraping technique could accomplish. Thousands of additional notices appeared online the next day, and most eluded Geilhufe's machine-readable XML format. A typical bulletin board message might read: "My father, Joe, was working in New Orleans and hadn't evacuated. He was living in Jefferson Parish.

We don't know if he's okay. Please call me or Mom in Houston. Lisa Brown, Houston, TX."

So that morning Geilhufe recruited a number of colleagues to help co-ordinate a massive data-coding effort. Jon Lebkowsky, cofounder of an in-formation management firm, Polycot Consulting, recruited volunteers to sift through all the missing persons posts. Ethan Zuckerman, a fellow at Harvard Law School, built a wiki to help dole out chunks of data to be parsed.

By Sunday morning, news of the PeopleFinder effort was spreading like wildfire across the blogosphere. High-profile bloggers called for volun-teers to assist the process, and many readers obliged. By the end of the day thousands were volunteering, and in total some three thousand are thought to have contributed. The flood of activity occasionally overloaded the makeshift databases, until management software vendor salesforce.com stepped in with a more robust back end.

By Monday evening 50,000 entries had been processed, and the num-ber continued to increase significantly, eventually reaching 650,000. Mean-while, people looking for friends or relatives could enter a name, a zip code, or an address into a search tool hosted on www.katrinalist.net to get an instant list of names matching their query. Over 1 million such searches were conducted in the immediate aftermath of the hurricane.[2]

Tales of heroic volunteer efforts are not particularly unusual. Disas-ters of this magnitude tend to bring out humanity's finer traits. What is re-markable is that the PeopleFinder project might have taken a government agency with loads of money a year or more to execute. Yet the Peo-pleFinder group rallied to pull it together in four days with absolutely no cost to the taxpayer. Mass collaboration at its finest.

Yes, PeopleFinder was a back-of-the-envelope solution that deployed quite rudimentary techniques when a more sophisticated software-coding solution might have automated the process even further. And yes, an agreed upon standard for collecting and sharing data from the get-go would have enabled all relief-oriented sites to interoperate, and organiza-tions like FEMA, Red Cross, and Google could have teamed up to initiate powerful Web services from the moment Katrina appeared on the radar. But in the absence of such standards, the overriding lesson is this: Given an open platform and a complement of simple tools, ordinary people can create effective new information services that are more resilient than bureaucratic channels. As Ethan Zuckerman said, "The goal was not to overengineer our tools for the data-entry effort, but to build something very quickly that would let people lend a hand. The solution we came up with was adequate to let three thousand people participate. And three thousand people, lightly coordinated, can do impressive things."[3]

PLATFORMS FOR WEB SERVICES AND COMMUNITIES

The notion that innovation proceeds through the recombination of existing ideas to form something new is not unique to the Web, or even the last century. In fact, it was Isaac Newton who famously said in a letter dated February 5, 1675, "If I have seen a little further it is by standing on the shoulders of giants." His modest explanation for how he achieved such incredible insight into natural phenomena has come to represent the idea that all innovations are ultimately cumulative, with each generation of advances resting on the previous.

Today, with open platforms for innovation inviting unprecedented participation in value creation, cumulative innovation is going into overdrive. Growing numbers of professional and amateur developers are creating their own content and applications by combining various fragments they find freely scattered across the Web. As described in Chapter 2, this fluid, combinatorial approach to innovation is making the Web look increasingly like a traditional librarian's nightmare—a noisy library full of chatty components that interact and communicate with one another. Techies call these components "Web services": applications that interact with other Web applications for the purpose of exchanging data.

In Chapter 5 we discussed how a new breed of listener-artists remixes music online to create new mashup singles and albums. Web services mashups are created according to the same principle: a programmer mixes together at least two services or applications from different Web sites to create something new, and often better than the sum of its parts. With a growing number of companies now exposing their APIs, the mashup phenomenon is taking off. Indeed, so long as the Web remains open, innovation will proceed in a spontaneous fashion as interlocking services and components are constantly remixed and improved by anyone with the skills and inclination.

Though mashups have the sheen of a hacker revolution, the truth is that many of these developments feed directly into the innovation strategies of the new Web conglomerates like Amazon, eBay, Google, and Yahoo. Understanding where they came from, and the future they are pointing toward, is vital in trying to decipher the competitive dynamics of the new Web. More than that, it is a signpost for how value is being created throughout the economy and the lessons for people and organizations tied to other sectors are abundant.

Companies like Google open up the APIs to their platforms to harness external ideas, talent, and energies on a mass scale. By doing so they leverage resources far beyond what they could afford to deploy internally, and they develop innovations much faster than internal innovation models allow. What's more, the innovations are typically serendipitous in nature, yielding the kind of unanticipated results that only a uniquely qualified mind is capable of producing. "We expect new and creative ideas to come out of this that we haven't thought of yet," says Google product manager Bret Taylor.[4]

We opened this chapter with a few examples of the applications and mashups that external developers are building on the Google Maps platform. Before the company decided to open up Google Maps, a number of hackers had already reverse-engineered the application to build their own services. Google was surprised by the ingenuity of their work (housingmaps and Chicagocrime, which maps Chicago's criminal activity, were two early nonsanctioned examples), so it decided to encourage more hacking by opening the APIs. When the APIs were officially released, developers started creating new applications at a frenetic pace, blending Google Maps with various data sources to produce interesting new combinations.

Platform Dilemmas

In their current incarnation, however, mashups present a thorny problem— they provide pretty poor long-term incentives for innovators, and most lack protection for data owners. To illustrate, it's worth taking a closer look at housingmaps.

As noted, housingmaps has two key ingredients: a Google Map and the rental listings from craigslist. When Paul Rademacher mashed these services together he created something new—something neither Google nor craigslist had thought of, yet it was a clever and useful application. In early 2005, Rademacher's application was generating a lot of buzz. It was a high-potential, high-demand application. What did Rademacher do to leverage this popularity? He stepped back from it and took a job at Google.

When asked why he did not develop the tool further, and perhaps even create his own business, he offered two very salient points: 1) He did not own the data that was powering the application; and 2) The barriers to re-creating such an application were low, since his Web site contained little in the way of unique intellectual property or user interface design, apart from a little bit of software code.

As housingmaps grew in popularity, there was a risk that Rademacher might face potential legal action from craigslist for appropriating its data, or that craigslist might simply opt to replicate the application itself. If craigslist failed to do either, there was no guarantee that copycat competitors would not simply leverage the same open API to create a knockoff service—the commoditizer gets commoditized.

Indeed, this highlights the problem facing craigslist. Housingmaps and similar mashups put craigslist's business model at risk. Craig Newmark's "classified ads community" allows advertisers to run classified ads in categories such as rentals, products for sale, job listings, and personals in major cities around the world. Over 99 percent of the site's content is free, but Newmark charges for job ads. Revenue from job listings in New York, Los Angeles, and California are estimated at more than $20 million per year—with very healthy margins for this tiny private company.

Apartment searches on housingmaps may not appear to pose much of a threat to craigslist's key revenue stream (the apartment ads are free, after all). But left unchecked, Newmark could watch housingmaps and other sites harvest his content in order to create superior services that attract users to other destinations. In turn, fewer and fewer eyeballs would make their way to those lucrative job ads, and this section itself could be pushed down a similar path.

There is no reason for craigslist to take this risk. Ownership of the data and its significant user base provide a sufficient barrier to entry. If housingmaps (or another Google Map/craigslist mashup) proved popular, Newmark could simply integrate the free API himself with minimal investment. Rademacher alluded to this. He had no real protection. And if he didn't have protection, what incentive did he have to pour more time and energy into his housingmaps application? Very little, it seems.

If Rademacher lacks long-term incentives, and if mashups are potentially threatening to craigslist, why is Google so fond of them? For one, Google's spectacular growth from the late entrant into the Internet search game to a dominant global company rests, in large part, on an open approach to innovation. Google was delighted with the housingmaps application—it was free publicity and free product prototyping. And in Paul Rademacher they discovered promising new talent, whom Google promptly hired.

Additionally, there are some unique elements to Google's business that position it to aggressively exploit platforms for participation. Search and mapping functions have a common characteristic—they help people find things. For these tools to be useful and profitable, they must be highly visible to Web users. Mashups increase Google's visibility by spreading Google Maps around the Web.

With online search, Google makes this visibility pay dividends by running advertisements alongside its search results. Each transaction is worth somewhere between a few pennies and a few dollars. Multiplied millions or billions of times over, this has proven to be extremely profitable. The online mapping applications can leverage the same model, and Google is reserving the right to run advertisements on any Google Map applications.

Google's opportunity to leverage the advertising-supported business model will only grow as the Web goes mobile. The combination of maps and search technology will serve as a key linkage point between the physical and virtual worlds. By having a piece of the customer-facing context, they may be able to extract value from an ever-increasing share of global transactions, a few pennies at a time. And what better way to find out what the killer applications will be than having large, self-organizing webs of developers building prototypes?

Of course, harnessing self-organizing developer ecosystems is by no means the only way to fuse applications and data sources. An alternative is "proprietary integration," whereby Web companies merge or integrate their applications through contractual relationships. But top-down contractual approaches are less fluid, less scalable, and lack the serendipitous nature of unplanned innovation that can surface nonobvious applications and services. Instead, Google sits back and lets self-organization take its course, confident that it will reap the long-term rewards no matter how its applications get integrated into the new Web tapestry.

Platforms Go Mainstream

That platforms for participation are beginning to permeate the mainstream is perhaps exemplified by the British public broadcaster BBC's recent launch of its own Web services initiatives. Through its Backstage project, the BBC invited developers to create new prototype services built around BBC content feeds like news, weather, and traffic. By harnessing external ideas and energy, the BBC hopes it can develop innovative offerings, like new ways to search and navigate BBC content, and perhaps even new revenue streams (though at the moment all Web services are strictly for noncommercial use). Specific communities of interest, after all, are likely to develop custom interfaces to BBC content that its internal developers might never have thought of.

Prototypes for hundreds of new services have been posted to the Backstage site since the project's launch in June 2005. Prototypes currently include the ingenious Mighty TV, which combines search, tagging, ratings, and peer recommendations to help viewers navigate their way through thousands of hours of UK television and radio. Alternatively, there is the more mundane but plainly useful BBC TV/radio/traffic guide for your cellphone.

A sister BBC initiative called the Creative Archive is opening up portions of the vast BBC content archive—which includes the largest television library in the world. The public is free to use this content as they like, again for noncommercial purposes. "Up until now this huge resource has remained locked up, inaccessible to the public because there hasn't been an effective mechanism for distribution," said Greg Dyke, the BBC's former director. "With the digital revolution, there is an easy and affordable way of making this treasure trove of BBC content available to all."[5] As an experiment, the BBC has run contests where participants are encouraged to remix BBC content (ranging from news reports to coverage of sporting events) into new media creations. Participants are then free to share their masterpieces with anyone they like.

What's encouraging about this initiative is that a large, mature organization like the BBC now recognizes that it is sitting on a deep wealth of content and a broad media platform that spans across the Web, radio, and television—a platform that will only become more valuable as members of the community build on it. Other media organizations would do well to follow the BBC's example. By hoarding content and capabilities—often for no better reason than cultural inertia and legal nuisances—they are missing significant opportunities. They could create dynamic open platforms, as the BBC has done, where large communities of users and developers participate in value creation.

The participatory Web services developed by Google and the BBC show the extent to which the ability to harness open platforms to drive collaboration and value creation is rapidly expanding. Still lagging in many of these examples, however, are strong business models and incentive frameworks that encourage and reward participation. Many APIs are available for noncommercial use only, and many mashups lack serious barriers to entry for their creators. Because of this, many Web services languish in beta form, lacking the resources or incentives to develop further. The companies that figure out how to harness the power of open platforms while providing adequate incentives to all stakeholders are poised to reap great rewards.

A little later in this chapter we examine some options for moving beyond "the culture of generosity" to a viable incentive framework for open platforms. One company already ahead of the game in this respect is Amazon, to which we now turn.

PLATFORMS FOR COMMERCE

One of the things we've learned through our research is that retail sites such as Amazon, eBay, and Apple demonstrate how platforms for participation can give rise to vibrant ecosystems around a simple activity like shopping.

For example, on eBay, high levels of peer-to-peer interaction are enabled via messaging, chat, and other applications. Social networks form as buyers and sellers with a particular passion, such as antique collecting, rare comic books, or vintage musical instruments, come together to share information as well as transact. Customers frequently use the simplest of tools to create a quick collaborative community for their own ends.

The communities at Apple, eBay, or Amazon also extend to "associate" stores: sites that sell their own wares using Amazon's interface and payment processing, or that sell Amazon's books and Apple's music downloads on their own sites. Vendors who integrate with a large firm in this fashion become important stakeholders. As eBay CEO Meg Whitman says, "We have a partner in this business and that partner is the community of users."[6]

At the highest level of integration are "developer ecosystems" of partner/developers. By using developer APIs, outside firms can connect to the database of products on sites such as eBay or Amazon and present them in new ways to customers. They extend the basic product by using it as a platform for development. In addition to merchants who sell their goods via these sites, other third-party software developers create tools for eBay and Amazon. Abidia makes software so users can track eBay auctions from anywhere. Other firms make software that enables Amazon shoppers to comparison shop from their cell phones.

With 1.9 million active seller accounts, thousands of developers, and third-party sales generating roughly 28 percent of Amazon's revenue there is arguably no company in business today that knows how to harness a platform for participation like Amazon, so it's worthwhile taking a deeper look at their model.

Amazon's platform for participation spurs two things: innovation and viral growth. Let's discuss innovation first. Most companies spend hundreds of millions of dollars on R&D every year with no guarantee that it will lead to the next great innovation. Amazon leverages a massive community of developers and small and medium-size businesses to delve into uncharted areas where traditional R&D models typically fail. Even better, Amazon incurs very few costs and risks—most of which are borne by the outside developers who create the innovations. Apart from the cost of maintaining the Web services, it's a virtually free development model where both parties win when developer creations increase sales.

How does Amazon do this? As in our previous examples, Amazon has opened up its APIs to its e-commerce engine in order to invite external participants to become codevelopers on the Amazon platform. Now external developers build ingenious applications ranging from Web sites that organize Amazon's catalog of CDs according to the top songs in rotation at major radio stations to an instant-messaging application that enables MSN and AOL users to ping an Amazon bot with a request and have it message you back with links to relevant products.

Why would developers agree to this? Simple: Amazon is the biggest game in town, and for a software developer they make a great customer. Amazon's Web services give developers access to numerous Amazon software services (like its shopping cart) and every fragment of data they could wish for (including text for product descriptions and reviews, product images, and pricing information). Amazon's Web services evangelist, Jeff Barr, calls these "the functional building blocks that external developers use to construct applications."

With functional building blocks in hand, Amazon gives developers carte blanche to build any application they see fit. No one has to ask for permission or await approval. There's no haggling over specs or schedules. It's simply: let a thousand flowers bloom.

For developers, the APIs are either free or inexpensive to use (in some cases, Amazon charges for data transfer or data storage). Moreover, developers stand to make good money from commissions on traffic and sales driven through their applications. Commissions on even a small percentage of Amazon's $25 billion in sales would yield healthy revenues for small developers.

By leveraging Web services for innovation, Amazon has gained a valuable head start in a wide variety of areas such as pricing transparency, RSS advertising, and comparison shopping. RefundPlease, for example, uses Amazon's own pricing information to inform customers if the price of an item they have purchased drops within thirty days of purchase. Amazon will refund the difference to the customer in a bid to boost loyalty.

FeedBurner (a popular news syndication service) and Amazon built a service that enables publishers to earn Amazon referral fees from the product links they embed into their newsfeeds. Every time a reader clicks on an ad in your blog feed, and that customer makes a purchase on Amazon, you get a sales commission—which is not a bad way to make a return on blogging. You'll have to split the commission with FeedBurner—but the tool was their idea, after all.

For those who still like shopping in the physical world but don't want to miss out on better online deals, there's ScanZOOM. This service enables users to snap a picture of a product bar code with their camera phone while they shop and receive instant price comparisons and product information on comparable items on Amazon.

Innovative services like these are unlikely to have emerged as quickly (or even at all) had Amazon opted to conduct all of its R&D internally.[7] But with its army of external developers and partners working to create new and innovative uses for the Amazon platform, Amazon has become one of the liveliest and most versatile business infrastructures around.

One might think that Amazon would want to closely guard all of its proprietary tools and data. But, in fact, the opposite is true. Barr says, "The more data that we're able to put in the hands of developers, the more interesting tools, sites, applications will be built, and the more of those that exist, the greater the return to Amazon. We're going to see more traffic, more clicks, and ultimately we'll see more purchases. So it's definitely not like a science experiment." Which leads us to viral growth.

Amazon is a pioneer in what are known as "affiliate programs" that it uses to drive traffic and sales through an immense network of external partners. Amazon has two principal types of affiliates: Amazon associates and marketplace sellers.

Hundreds of thousands of Amazon associates send traffic and sales to Amazon from their own Web sites and get paid commissions for doing so. It's akin to creating a customized front door to Amazon's product catalog, with each door offering something unique to customers. Many associates drive sales through Web links and advertising. But the more sophisticated associates are harnessing Amazon's payment and distribution infrastructure to set up their own specialty stores that sell products from the Amazon catalog. A number of software vendors offer "associate-o-matic" programs that enable virtual retailers to get up and running in about thirty minutes. Now thousands of niche associate stores are flourishing by selling everything from power tools to books on west Texas.

Hiking Outpost is a specialist book retailer that sells thousands of books about (you guessed it) hiking. A customer visiting its site might not realize it at first, but all of the product descriptions, customer reviews, comparison pages, images, and even the search function are seamlessly provided by Amazon. Hiking Outpost fosters its own identity and offers a superior experience to customers by aggregating valuable information about camping sites and hiking trails from all over the Web. Amazon handles all the credit-card processing, order shipping, and returns for products sold on the site. Like other associates, referral fees for Hiking Outpost range from 4 to 7.5 percent, and are based on the total number of items shipped from both Amazon and the third-party sellers. So Hiking Post makes a small, but reasonable profit on every item they sell for Amazon (which is not bad considering that Amazon does all of the complicated stuff like order processing, data management, and distribution) while Amazon increases its revenue and growth.

Amazon marketplace sellers are different from associates in that they maintain their own inventories of products (perhaps books, DVDs, or CDs) that they then list, sell, and distribute through Amazon. Over a million registered sellers, ranging from merchants to individuals, now use the Amazon platform to sell new and used goods.

All of this adds up to an expansive and fertile ecosystem of developers, associates, and sellers who are fueling rapid growth for Amazon. Barr says associates and marketplace sellers are "increasing the surface area of Amazon." They add more and more things to sell, in more and more places on the Web. All of this happens in a completely self-organizing fashion. And this makes Amazon's already low overhead even lower.

Saving the best for last, Amazon (through subsidiary Alexa) is leveraging a Web services approach in a bold attempt to bolster its presence in the online search business. Amazon launched its feature-rich A9 search engine to much fanfare in April 2004, but capturing less than 5 percent of the market today it languishes well behind search leaders such as Google, Yahoo, and MSN. Rather than throw in the towel, Amazon has changed the game completely. It's offering its proprietary Web index and search technology up to anyone who wants it—for ridiculously low prices. In doing so, Amazon has opened up the search industry in the same way that open source democratized desktop software. In the meantime, the company has challenged all of the latent programming talent in the world to help it build a better search engine.

How does this work? Amazon is renting access to the whole raw database of roughly ten billion Web documents so that anyone with the time or inclination can build their own search tools and data-mining projects with it. Indexes are hard to build and maintain, requiring a lot of computer horsepower, storage, and bandwidth. But once you've built a copy of the Web index, there's plenty of imaginative ways you can tweak it to produce valuable new applications and services, and perhaps entirely new search engines.

Again, it's all done via Web services. There are no licensing fees for developers. Just "consumption fees," which, according to Jeff Barr, are pretty reasonable: one dollar per CPU hour consumed, one dollar per gig of storage used, one dollar per fifty gigs of data processed, and one dollar per gig of data uploaded if you are putting your new service up on their platform.

By treating the index as a salable asset instead of a trade secret, Amazon is really ripping apart the traditional wisdom of search engines. The Alexa index is now a platform for participation that anyone can harness without investing millions of dollars in crawl, storage, processing, search, and server technology. In doing so Amazon effectively commoditizes the technology deployed by the big guns in search (Google, Yahoo, and Microsoft), while monetizing it for itself. No doubt the big guns are sitting up and taking notice.

Amazon's open platform approach brings many unique advantages. First, it innovates faster than competitors and stays at the forefront of innovation by leveraging external resources and talent to push the boundaries of its technology and applications. Amazon harnesses the strength and breadth of its developer ecosystem to release updates to its platform frequently and builds in powerful feedback loops that enable it to respond to problems in weeks, not months.

Second, by opening up the APIs to Alexa and its e-commerce engine, Amazon has essentially turned its platforms into salable products. Some might argue that this commoditizes the software that gave it a competitive advantage in the first place, thus eroding Amazon's competitiveness. On the contrary, Amazon enhances its competitiveness because there is now less incentive for competitors (especially smaller rivals) to build a competing platform when they can leverage Amazon's for free. The old "if you can't beat 'em, join 'em" kicks into gear.

Third, Amazon's Web services program is based strongly around reciprocal benefits for all participants. Amazon perpetuates its brand, increases revenues, and gains valuable IP from outside its corporate walls. Developers and partners gain access to state-of-the-art software and earn sales commissions—it's a win-win scenario.

Finally, Amazon got way ahead of the game by starting its Web services program in 2002. It has solidified its lead with an active evangelism program that offers technical support, transfers knowledge across the community, and makes its developers and affiliates feel like genuine stakeholders. Amazon is already the world's largest online retailer. As it leverages its ecosystem to gobble up more surface area on the Web, the company is poised to become the dominant retail force on the planet, period. If we were running Wal-Mart, we would be very scared indeed.

PLATFORMS FOR GRASSROOTS ACTION

Tech firms like Amazon, eBay, and Google may have pioneered the use of APIs to unleash a torrent of business innovation, but the potential for widespread application of this approach is surely limitless. A sure sign that the approach is catching on is that a small number of government agencies are getting on the API bandwagon. This is an opportunity whose time is long overdue. In 2000, we led an international research program aimed at moving government and governance into the digital era. The program produced many valuable initiatives, many of them ahead of their time. But scores of government leaders struggled with cultural inertia, complex legacies, and political wrangling. As the rest of the world raced ahead, governments were being left behind.

In the six years since the program concluded, there have been large strides forward in modernizing government service delivery. But the potential for public sector innovation has barely been tapped. As Jim Willis, director of eGovernment for the office of the secretary of state for Rhode Island, observed, "It is simply unacceptable at this point in history that a citizen can use Web services to track the movies he is renting, the weather around his house, and the books he's recently purchased but cannot as easily monitor data regarding the quality of his drinking water, legislation, or regulations that will directly impact his work or personal life, what contracts are currently available to bid on for his state, or what crimes have recently occurred on his street."[8]

He's right. Government agencies are one of the largest sources of public data, and yet most of it goes completely unutilized, when it could provide a platform for countless new public services. Both the private sector and advocacy groups like Greenpeace are much farther ahead in using new technologies to disseminate and leverage information to empower their operations.

Secretary Brown believes that the gap can be closed. Under his leadership, the state of Rhode Island recently began publishing large amounts of government data via an API that is easy to hack. Brown hopes his Gov-Tracker application will at least bring Rhode Island a step closer toward an era of community-developed applications that will make it as easy for citizens to interact with their government as it is for them to interact with the rest of the networked world.

The truth is that the best uses of public data are often made by organizations in the nonprofit sector that are free of the political considerations that hamstring government agencies. Governments should move faster to create new platforms for participation and public knowledge. A good start would be making more public information accessible to people and organizations that could put it to productive uses. In the meantime there are a number of great examples for government officials to draw from.

Scorecard, for example, might just be the mother of all mashups. The Environmental Defense Fund (EDF) launched the application in 1998 (that's right, a full seven years before the term Web 2.0 was even invented) to aggregate hundreds of sources of public data to create a powerful nationwide tool for assessing environmental risks. Then there's Neighborhood Knowledge California (NKCA), an ingenious tool that harnesses public data to help citizens and policy makers spot and improve troubled neighborhoods. It was created by researchers at UCLA who joined up with community groups in Los Angeles to empower low-income neighborhoods like Vernon-Central to reclaim and rebuild their neighborhoods.

These and other strategies should become part of a more concerted effort by governments to explore and leverage new forms of value from public information. To some extent, the sophistication of these efforts gives even the most dyed-in-the-wool Web 2.0 aficionados something to learn. For a window into these insights, let's look at Scorecard and NKCA in turn.

Platforms for Public Disclosure

One look at Scorecard and it's easy to see why it's every industrial polluter's worst nightmare. But to fully appreciate the significance of Scorecard, it's worth briefly retracing its history. The story begins with a tragedy in Bhopal, India, where a chemical explosion in a Union Carbide plant in 1984 killed over 3,500 people and acutely injured tens of thousands of others. As public awareness of the dangers posed by toxic chemicals grew—spurred by domestic events like Love Canal—policy makers came under intense pressure to protect the American public from a similar public-health disaster. Community leaders from across the country became increasingly vocal in demanding access to information about nearby environmental hazards.

In 1985, members of the U.S. Congress responded with a bold new plan to inform the public and put the spotlight on polluters. They drafted the Emergency Planning & Community Right to Know Act, which contained a provision called the Toxic Release Inventory (TRI) that empowered the Environmental Protection Agency (EPA) to collect emissions levels on 328 deadly chemicals in use in commerce. With considerable foresight, Congress required the TRI data to be made available to citizens via computers (of course there was no public Internet then).

The bill met furious opposition from industry, and even from some EPA officials, but narrowly passed. The first report, released in 1989, showed that billions of tons of toxic waste were being released into the environment. Observers credit this report with spurring the chemical industry to intensify the search for low-pollution technologies.

Around the same time, environmental groups were gaining valuable new ammunition to deploy in their war against industrial polluters. Geographic information systems (GIS), Web, and computer simulations, for example, gave them the ability to collect, manage, and distribute large volumes of environmental data in a way that only elite government agencies could manage in the past. When the Internet came online in the early 1990s, it provided everyday citizens with the most powerful platform ever to find out, inform others, and organize.

Environmental groups soon recognized the enormous power of the Internet and began building Web-enabled systems to harness the TRI. Of the many that have emerged, Environmental Defense Fund's Scorecard is by far the most sophisticated. Scorecard combines data from over four hundred different scientific and government databases to profile local environmental problems and the health effects of toxic chemicals, making it one of the most advanced sites on the Web in terms of informatics.

Visitors to the site can type in their zip code and get instant access to a wealth of information about pollution sources in their region. Want to know, for example, which company is the biggest source of air pollution in the state of California? Scorecard's database says it's an ExxonMobil refinery located at 3700 West 190 Street, in Torrance, California, with an annual release of 1,659,872 pounds of toxic stew. Or perhaps, as you plan that next real estate purchase, you care to avoid the country's most polluted communities. Scorecard says you better sidestep Humboldt County, Nevada, where a staggering 350 million pounds of carcinogens are released annually.

It took more than a year to develop Scorecard, with more than a million dollars worth of programming time—much of it donated. Its immediate popularity surprised the EDF—the site received more than a million hits in its first two days in April 1998. Today interested users can access Scorecard from over five thousand community portals, municipal Web sites, real estate brokers, and the homepages of countless environmental organizations.

Scorecard's strengths lie in its powerful but highly accessible interface to complex datasets. One innovative application, called the Pollution Locator, harnesses a Web-based map server to generate environmental data charts dynamically, as the user's cursor crosses geographic areas. Users can zoom out to compare data between counties or states, or zoom in to examine neighborhood environmental problems at the street level.

The creators of Scorecard built an easy-to-use tutorial that steers users to information about environmental hazards in their community. The site offers lay explanations about each type of pollutant it tracks and their associated health effects—turning raw information from TRI and other sources into practical knowledge. Visitors to the site can also personalize the way Scorecard displays information, and send e-mails and faxes directly to polluters. For those inclined to organize communitywide action, it has an online forum where concerned citizens post questions, give advice, find other concerned people in their community, and network with those who have had similar experiences. A list of polluters even includes the plant supervisor's phone number.

Platforms for Neighborhood Knowledge

While Scorecard is largely about naming and shaming polluters, Neighborhood Knowledge California (NKCA) is empowering residents to improve their communities. The cornerstone of the project is an online tool that provides easy access to a vast collection of public data about properties and neighborhoods facing urban decay. Here's the dilemma. Spotted early enough, a community's decay could be reversed through a combination of well-targeted public programs and private sector investment. But although the danger signals are all on public record, they are effectively inaccessible to the public, buried deep within the bowels of city hall. NKCA illustrates what can be done when simple Web-based tools transform raw public data into formats that are meaningful and useful to community residents and local-government policy makers.

The NKCA project integrates databases containing information about public (city, county, state, federal) and private (i.e., investment, toxic release notices) activities that can be tracked at the neighborhood level to develop an interactive Neighborhood Electronic Monitoring System (NEMS). NKCA's evolving information system uses a mapping interface to plot near real-time information on city maps posted on the Web site. Rather than having to look at each database separately, public officials, citizens, and businesses can search by zip code or other parameters to view comprehensive information on one property, or see at a glance which communities might be headed for trouble.

NKCA was first developed at UCLA's Center for Neighborhood Knowledge as a participatory research project with residents from Vernon-Central, a low-income neighborhood in Los Angeles.[9] While the researchers were seeking to better understand the patterns and processes of residential disinvestments, they wanted their interactive research tool to empower community members to spot early warning signs that properties in Los Angeles are headed for unlivable status. Since NKCA launched in September 1999, citizens and community organizations have been using its online databases to look for properties with tax problems, code violations, or other difficulties, such as tenant complaints or fire violations, that could be precursors to abandonment and deterioration in their neighborhood.

The focus of the project is not just on identifying problems. NKCA has developed a code-enforcement tracking system that enables residents to monitor Los Angeles's responses to housing-code complaints and violations—similar to the way online customers track their FedEx packages. Users are supplied with information on how to conduct their own inspections, contact city inspectors using electronic forms for complaint letters and other documents, and find assistance in resolving housing concerns, including mediation groups. NKCA researchers also work with grassroots community organizations, tenant groups, and activists to promote code enforcement by government officials. These grassroots efforts, in turn, are helping improve compliance by property owners.

One early criticism of NKCA was that it often reinforced the image that low-income communities only contain "deficits," such as nuisance properties and environmental hazards. Moreover, the information displayed by NKCA was supplied by government databases, while the real neighborhood experts—i.e., the people who actually live there—were initially not invited to contribute their own information. Those issues have since been redressed, and today, with minimal expertise, community members can upload their own spreadsheets into NKCA's system and create their own personalized maps. Youth in the Vernon-Central community, for example, have initiated an electronic "treasure hunt" where they are using GIS-enabled devices to locate and describe spaces important to them and their community. Now residents can use NKCA to find information about church groups, organizations, social programs, and youth activities. It's all part of a wider community-asset mapping initiative run by groups like

Concerned Citizens of South-Central Los Angeles, in which NKCA is playing a key role in helping community members identify strengths for rebuilding.

NKCA, Scorecard, and similar grassroots projects are great examples of how platforms for participation that empower more people to become involved in identifying and resolving problems in their communities can improve public sector governance and enrich democracy. For communities left out of the high-tech boom in particular, open platforms and well-designed Web services can provide real hope when they are applied to concrete social problems. Community outreach, access to technology, and training help low-income and linguistically isolated communities connect to effect social change. In fact, the powerful combination of interactive mapping applications and citizen participation could easily be replicated to track information on issues such as employment, public health, and migration patterns.

Now if local governments and nonprofits can get their heads around the power of open APIs and Web services, can you imagine what might be next?

PLATFORM INCENTIVE SYSTEMS: BEYOND THE CULTURE OF GENEROSITY

A blog entry from Anil Dash, a vice president with Six Apart (a maker of blogging software), dated October 25, 2005, started an interesting debate: Should flickr compensate the creators of the most popular pictures on its site? The premise is simple: Flickr trades free hosting of pictures for the right to post ads. The most popular and/or best pictures draw the most clicks, in turn creating a disproportionate amount of value for flickr. Compensating the creators of the most popular content might improve the quality of pictures (thus driving traffic) and allow people to be compensated for their work: basic capitalism at work.

It's a pivotal question: Should open-platform orchestrators compensate the people and organizations that add value to their platforms? And would monetary incentive systems spur more value creation or possibly taint the dynamics that have made online communities like flickr successful? Companies that have been in the platform business for some time, like

Amazon, have figured this out. Amazon's revenue-sharing agreements have enabled many, many new businesses to flourish. But many new Web companies have not matured to the point where they have established sustainable grounds rules for their ecosystems.

Caterina Fake, a cofounder of flickr, responded to Anil's query by saying that there are systems of value other than, or in addition to, money, that are very important to people: connecting with other people, creating an online identity, expressing oneself—and, not least, garnering other people's attention. She continued on, saying that the Web—indeed the world—would be a much poorer place without the collective generosity of its contributors. The culture of generosity is the very backbone of the Internet.

There certainly is truth to that statement. People motivated by non-monetary rewards have driven a lot of innovation—just look at Wikipedia, open source software, and, indeed flickr. But might the multimillion-dollar maneuvers of Google, Yahoo, and, Microsoft trample the spirit of generosity that has sustained this run of free-spirited innovation?

While the culture of generosity and expression may have contributed to the popularity of flickr, the founders did sell it to Yahoo for an estimated $30 million. Del.icio.us's founder, Joshua Schachter—anointed by many as a Web 2.0 leader—followed the exact same path by selling out to Yahoo in December 2005. While craigslist is an "open network," Craig Newmark profits quite handsomely. Google's propensity to hire people who do interesting things to its code (such as Paul Rademacher) that spur its rapid growth certainly helps them attract development talent—and helps generate a war chest of advertising revenue. But could the culture of generosity amount to little more than a smoke screen on what ultimately amounts to an exploitative phenomenon?

Om Malik, a well-read blogger and founder of GigaOmniMedia, points squarely at this dilemma in a recent blog post:

> *I wondered out loud if this culture of participation was seemingly help[ing] build businesses on our collective backs. So if we tag, bookmark, or share, and help del.icio.us or Technorati or Yahoo become better commercial entities, aren't we seemingly commoditizing our most valuable asset—time. We become the outsourced workforce, the collective, though it is still unclear*

what is the pay-off. While we may (or may not) gain something from the collective efforts, the odds are whatever "the collective efforts" are, they are going to boost the economic value of those entities. Will we share in their upside? Not likely!

Calling it exploitation goes too far. But like the early days of Web 1.0, the business model issues have not kept pace with the speed of innovation.

However, this time the dominant online players—Google, Microsoft, Amazon, Yahoo, and eBay—are well-established, profitable, multibillion-dollar enterprises. Collectively they acquire the vast majority of interesting applications that emerge online. The competitive dynamics between these entities are evolving rapidly as they increasingly converge on common competitive spaces, helping people find things, facilitating commerce, and enabling community interaction. They need to be careful not to violate community norms. At the same time they are seeking to leverage their platforms for competitive advantage. The winners in this evolution will be companies that can create the most comprehensive incentive frameworks to adequately reward all stakeholders.

Google's decision to open up its mapping application stimulated a new wave of Web services. Unfortunately, the most interesting applications tend to infringe on the rights of other stakeholders. Noninfringing services are currently limited to integrating mapping applications with nonproprietary data sources, such as the public data sets used to fuel applications like Scorecard and NKCA. As valuable as these applications are to citizens and to the health of our democracy, they will not drive substantial economic value, because the barriers to imitation are low. A business model that protects the rights of craigslist while still providing incentives for people like Paul Rademacher, on the other hand, could define the eventual winners.

Some companies use contests to encourage users and developers to innovate on their platforms. EBay and Microsoft were simultaneously running such contests at the time of this writing, with prizes ranging from $5,000 to a free Xbox. But these offerings are trivial, and perhaps even insulting, when hundreds of millions of dollars of online revenue are at stake. A more comprehensive framework might include royalty payments to top innovators. Or, as Google, Technorati, and others are currently doing, giving away free tools but reserving the right to take a stake if someone creates a going concern. The point is that there is a vast array of options between employing full-time development resources and an online community offering free services.

As noted earlier, Amazon is arguably ahead of everyone. Amazon's developer ecosystem and affiliate program are now legendary. But even Alexa's Web search offering brings an interesting new business model into the competitive search space, allowing innovative developers to leverage Alexa's Web index to create new search tools on a pay-per-use model. This development could bring a new competitive element to the Google/ Microsoft/Yahoo search battle. While the leaders fight over their "one size fits all" search engines, Alexa's Web services may lead to a customized suite of search solutions that have been developed for particular communities of interest. Both external developers and Amazon share in any upside.

Similar incentive frameworks could apply equally to the world of social networking. Indeed, there is no reason that the best photographers, taggers, and other online contributors can't share in the rewards being created by their work. Some devoted users of flickr and del.icio.us begrudge the fact that Yahoo is acquiring the communities they helped to build, and are understandably afraid Yahoo will poison the culture of these communities with unbridled mercantilism. They certainly have a point. Without their content, the community is about as valuable as a ghost town. If Yahoo can rake in millions in revenue, shouldn't the community members share in it?

Blogger and media consultant Jeff Jarvis (who we first met in Chapter 5), points out that even a simple act of consumption in this new world is now an act of creation. Acts like searching on Google, tagging bookmarks with del.icio.us, and sharing photos on flickr all have private benefits, but these acts create collective benefits as well. These collective benefits yield a richer Web experience and enhance "the wisdom of crowds." This new wisdom, says Jarvis, can be useful in helping people discover content, in organizing the Web around topics, in improving search results, and even in improving ad performance.[10]

"So who owns that collected wisdom of the crowd?" asks Jarvis.

Obviously, the crowd does. Platforms like Google, Technorati, and Yahoo (including their new subsidiaries flickr and del.icio.us) merely borrow it. And they can only borrow it, says Jarvis, "if they continue to have the trust of the crowd and if they pay dividends back to that crowd. And those who try too hard to control that wisdom, to limit its use and the sharing of it . . . risk turning away the crowd that creates this value."[11]

If Yahoo, for example, layered in a system that shared the rewards generated by the community, disgruntled community members might change their tune. What's more, a new era of online "micro business" could be born. Just as over a million people now make a living on eBay, we might see the rise of entrepreneurial del.icio.us taggers who are rewarded for directing people to useful content, or flickr artists who are commissioned to post great photos that keep the community coming back.

What do the platform providers deserve in exchange for enabling all of this activity? They should profit from it, and profit handsomely if they can. Most platform providers emphasize the need to build the largest network possible first, and they claim the profits will soon follow.[12] It sounds a little like dot-com logic, but the difference now is that you can build with less investment, provide an environment for experimentation, and then seize on the things that users find valuable. The key to this is openness, and in giving the users control and freedom. Put profits first, they say, and you will cripple the network you are building.

Companies that build openly and build as big and as fast as possible are ultimately in the best position to figure out where the real economic value is. Google built the world's most popular and useful search engine, and eventually became an ad company. Skype built a free phone service, and eventually sold it to eBay. Craigslist built a free classified-ads community, and is turning a healthy profit—while killing traditional publishers—with its comparatively low-cost job ads.

As the new Web evolves, platforms for participation are becoming the competitive standard for top online properties. There are a variety of ways to compete in this environment, and open approaches to innovation do not have to be synonymous with free. Companies that attract and reward the best participants have the opportunity to create new sources of competitive advantage.

WINNING WITH AN OPEN PLATFORM

Three key points stand out from our discussion of open platform business models. First, all applications of a new technology go through an evolutionary process in which a period of early experimentation gives way to shakeouts, and then the truly viable business models emerge. Second, radical decentralization and openness create tricky environments in which to build genuinely viable business models—success lies in "closing up" the right parameters and rewarding innovation without destroying the characteristics of the system that makes it innovative. Third, platforms for participation will only remain viable for as long as all the stakeholders are adequately and appropriately compensated for their contributions—don't expect a free ride forever. With these lessons in mind, we offer a few concluding thoughts.

Conventional wisdom says that being open is rather like inviting your competitor into your home only to have them steal your lunch. But in an economy where innovation is fast, fluid, and distributed, conventional wisdom is being challenged.

Winning in a world of cocreation and combinatorial innovation is all about building a loyal base of innovators that make your ecosystem stronger, more dynamic, and more expedient than the ecosystems of rivals in creating new value for customers. To achieve this, your organization—regardless of the sector or line of business—needs to identify and open up platforms to enable mass collaboration. That platform may be a product (e.g., a car or an iPod), a software module (e.g., Google Maps), a transaction engine (e.g., Amazon), a data set (e.g., Scorecard and NKCA), or countless other things we haven't covered here.

Giving away the keys to your most prized assets is not something you should take lightly. It's a bit like signing a free-trade agreement with the external world after a lifetime of protectionism. Shai Agassi, president of SAP's product and technology group, says, "It's almost like you're taking down your borders and opening up for no tariffs, no tax competition. You need to know that your core assets and your skill sets allow you to continue to innovate fast enough as a corporation."

SAP just recently went through the process of opening up thirty thousand APIs to its market-leading enterprise software platforms. "You need to decide, as a corporation, whether you take your core assets and processes and keep them to yourself, or do you expose them to every software company on the planet and entice them to come in and help develop those assets," says Agassi. "We believe that our strength, our genome, our understanding on how to build applications is significantly enhanced by this kind of a collaborative innovation marketplace."

Agassi does worry that competitors could come in and try to eat SAP's lunch. "But our customers love it," he says. "They get this collective innovation process—a large pool of innovative software companies can now provide them with additional solutions with integration by design, not integration as an afterthought." Agassi says SAP's ecosystem includes over half a million independent developers.

The key is that there are considerable advantages to be gained from acquiring network effects. And once a platform like Google, Amazon, or SAP gains traction there is less and less incentive for people to defect to other platforms. In fact, the trend is self-reinforcing. More developers create better offerings, whether products, experiences, or applications. Better offerings attract more customers. Growing customer bases attract more participants to the platform. And on the cycle goes, creating a dynamic succession of cocreation, innovation, and growth.

In an economy where growing numbers of individuals eke out a living as free agents, open platforms become all the more important. Agassi has a nice way of putting this. He says, "Most of the free electrons will gravitate toward the biggest centers of gravity." In other words, companies with the most dynamic platforms—and with great opportunities for partners to establish a synergistic business—will have the best chance of harnessing the enormous wealth of talent that all of those free agents can offer.

After all, success in most platform-related businesses is linked to pervasiveness and continuous innovation. The bigger the ecosystem the better, because bigger ecosystems support more raw intelligence and more requisite variety. Becoming a pervasive and continuously innovative presence means becoming a magnet for innovation that attracts lots of partners, suppliers, developers, customers, and other interested participants who are willing to build on your organization's platform.

Google, for one, doesn't care who innovates, and in some cases, it doesn't even care who controls the context for individual Web interactions: If

Google's applications are pervasive, they can profit regardless. Indeed, the more people developing applications on top of Google Maps and other tools, the better. It's like adding an army of R&D professionals to extend their platform, except they needn't add them to their payroll.

Platforms for participation represent an exciting new kind of business that thrives on mass collaboration and embodies all of the wikinomics principles: openness, peering, sharing, and acting globally. It's a business that many managers only dream of, where hundreds of thousands of partners work together in vibrant business ecosystems. Though the early examples are most evident on the Web, nearly all businesses can become open platforms with enough imagination and ingenuity.

8. THE GLOBAL PLANT FLOOR
Planetary Ecosystems for Designing
and Making Things

Somewhere in rural Ghana a team of students is working on low-cost designs for mobile refrigeration. They hope that one day soon their refrigerator designs will be manufactured in local villages across Africa, not by General Electric or some other multinational, but using a fabrication laboratory supplied by $25,000 worth of technology from MIT.

In a small remote village in India local villagers are using an identical fab lab to make replacement gears for out-of-date copying machines, reliable tools for testing milk contents, and diagnostic devices for human blood. In the Lyngen Alps of Norway, shepherds are keeping track of their flocks from afar, using fab lab–constructed wireless networking devices. Local fishermen are using the same technology to keep track of their boats at sea.

The all-purpose machine behind this local innovation and manufacturing was put together with off-the-shelf, industrial-grade fabrication and electronics tools, wrapped in open source software written by researchers at MIT's Center for Bits and Atoms. They call it a "fab lab," but think high-tech workstation meets assembly line. It has everything you need to make just about anything, including nifty gadgetry such as laser cutters to etch out 2D and 3D structures, digital carving tools for making circuit boards and other precision parts, and a suite of electronic components and programming tools for constructing cheap microcontrollers.

Putting a fab lab in every home, says MIT professor Neil Gershenfeld, would deliver a profound change in how we design and assemble physical goods. Just as the information revolution placed the means to manipulate information and media in the hands of everyone within reach of a computer, a similar wave of digital fabrication technology could eventually put the means to produce physical objects in the hands of every household and community. This, in turn, would radically transform the way we produce, consume, and interact with physical objects, perhaps even making us genuine producers of the everyday objects that have long been the province of large-scale industrial manufacturers.

In theory, it's true that many things people want could be manufactured right in the home to the exact specifications required. But whether having a personal factory in your home is at all practical, or even desirable, is debatable, to say the least. It's unclear, for example, how people would procure the raw materials for individualized production, or whether personal fabrication would ultimately be cheaper than today's mass-production systems. One certainty is that it will be many years before we even know whether personal fabrication of the kind envisioned by Gershenfeld is even feasible on a large scale.[1]

And yet, the idea that participation and collaboration are on the upswing in the design and assembly of physical things is not far out at all. In fact, we edge closer to a more collaborative reality every day as we design and develop physical goods in ever more decentralized networks of individuals and firms using methods that increasingly mirror those used to produce intangibles like knowledge. Soon the collaborative methods of open source software developers will be as amenable to cars and airplanes as they are to software and encyclopedias. All of this, in turn, is part of a momentous change in the manufacturing-intensive industries, as a truly global plant floor emerges to replace the traditional patchwork of national and regional production facilities. Peer production of physical things is coming of age, and smart companies are getting with the program.

RISE OF THE GLOBAL PLANT FLOOR

A key message in this book is that the old monolithic multinational that creates value in a closed hierarchical fashion is dead. Winning companies today have open and porous boundaries and compete by reaching outside their walls to harness external knowledge, resources, and capabilities. Even the stodgy, capital-intensive manufacturing industries are no exception to this rule. Indeed, there is no part of the economy where this opening and blurring of corporate boundaries has more revolutionary potential.

As long as humans have a need to eat, and to be mobile, housed, clothed, and healthy, physical things will be important to the economy. Now the companies that design and make these goods are beginning to adopt the four principles of wikinomics: openness, peering, sharing, and acting global. We're witnessing the rise of distributed networks that build and distribute goods—typically on a global basis. The plant floor is going global, and it's harnessing mass collaboration to design and assemble things more efficiently.

This is a sharp discontinuity from the reigning model of multinational production. The quintessential multinational was modeled on a hub-and-spoke architecture. A head office drew up plans and issued commands to an international network of satellite production facilities that built products for local markets. The corporation was a collection of subsidiaries, business units, and product lines, not a unified global operation.

Local production had and still has its advantages. Producing locally provided an opportunity to tailor products to local tastes. Hiring local talent created jobs and wealth in the local economy, which in turn boosted demand for consumer products. By avoiding international trade, companies escaped tariffs, exchange controls, and other trade barriers that came with the protectionism of the nation-state era.

This market-by-market approach to organizing production no longer makes sense in a global age. National silos gave rise to bloated and expensive bureaucracies that deployed inefficient, incompatible, and often redundant processes for making and marketing products locally. Insufficient knowledge transfer across organizational boundaries and departmental silos meant that most multinationals failed to seize opportunities for innovation and cost reduction. Now that global business standards and information technologies envelop the planet, the cost of coordinating a distributed global business organization is infinitely cheaper than just a few decades ago. Meanwhile, receding barriers to trade are freeing goods, knowledge, capital, and people to circulate according to their own market logic.

Companies that appreciate these changes are moving to a new model—a truly global firm that breaks down national silos, deploys resources and capabilities globally, and harnesses the power of human capital across borders and organizational boundaries. This is not an old multinational with a new twist. Smart firms are abandoning the multinational model completely.

In its place leaders are building globally integrated ecosystems that encompass hundreds, if not thousands, of firms. These new global enterprises assemble components of business activity and production on a global basis to produce goods and services for customers. Everything from conceiving the customer offering through to its delivery is loosely orchestrated in a seamless global collaboration.

It's also not simply a new spin on the old supply chain. Suppliers have growing power and increasingly play critical roles in everything from design and manufacturing to distribution and aftermarket repair services. Rather than thinking of them as "suppliers" it makes more sense for companies to view them more like partners, in some cases even peers.

Multibillion-dollar manufacturing specialists like Celestica, Jabil Circuit, Foxconn, Flextronics, and Solectron build computers, cell phones, video-game consoles, network routers, televisions, and other devices for pretty much everybody in the consumer electronics industry. But they are arguably more than suppliers. They contribute to product design, testing, distribution, and repair. They have each made significant investments to do this work. If they fail to have almost perfect quality their clients could be discredited, or even fail in the marketplace. The term "supply chain" was appropriate for the old hierarchical corporation, but it is not for the twenty-first-century firm. Today chains are becoming value networks.

In fact, the rise of planetary ecosystems for designing and building physical goods marks a new chapter in the corporation's evolution. As IBM chair and CEO Sam Palmisano recently put it, "The emerging globally integrated enterprise fashions its strategy, its management, and its operations in pursuit of a new goal: the integration of production and value delivery worldwide."[2] Not since the nineteenth century have our systems of production endured such large and fundamental changes to their architecture.[3]

As with all of the new business models discussed so far, the rise of a global plant floor presents tough choices for managers regarding how to mark the boundaries of the firm. When the membrane of an organization becomes porous, and companies network together to create value, how do you decide what should be in and what should be out?

The new reality in manufacturing, as in other spheres, is that boundaries are constantly blurring. Everything from an Apple iPod to an Airbus A380 to an Intel chip set combines components and services from multiple firms—often hundreds of them. In an age of modularity, open architectures, instant communication, and globally dispersed capabilities, the answers to who will do what and where will the value be created are constantly changing. All companies need an evolving sense of where their core capabilities lie and an evolving map of how they relate to the constellation of knowledge and capabilities that exist within their ecosystem.

We and our colleagues have argued for years that companies should treat their various functions and operations as component pieces that they can pull apart and recombine as necessary. Palmisano warns that "these decisions are not about simply a matter of off-loading noncore activities, nor are they mere labor arbitrage. They are about actively managing different operations, expertise, and capabilities so as to open the enterprise up in multiple ways, allowing it to connect intimately with partners, suppliers, and customers."[4] In other words, companies should base their boundary decisions on strategic judgments about which operations they want to excel at and which they think are best suited to partners, suppliers, and customers. In recent years this new imperative has yielded some interesting new developments.

A growing number of cars are not made by car companies anymore, at least not the companies most consumers recognize. BMW focuses on marketing, partnering, and customer relationships, and it maintains the engineering expertise it deems critical. But suppliers make most of the components, and increasingly they assemble the final vehicle. Specialization rules, and it turns out that a company like Magna International can assemble a vehicle faster, cheaper, and with better quality than BMW.

The same sort of thing applies in aerospace and defense. Modern aircraft consist of tens of thousands of high-tech parts. In the past companies like Boeing wrote detailed specifications for each part and asked suppliers to build to plan. Boeing gathered the parts on the plant floor and spent weeks assembling a single airplane. Today suppliers codesign airplanes from scratch and deliver complete subassemblies to Boeing's factory, where a single plane can be snapped together like Lego blocks in as little as three days.

Most people think of BMW and Boeing as seasoned innovators—renowned for harnessing their deep core of engineering expertise to bring industry-leading innovations to their respective markets. The fact that they are now handing significant responsibility for innovation over to suppliers signals an important change in how these companies compete. Developing and bringing new physical goods to market now means working with a vast ecosystem of partners that possess complementary skills and capabilities. For the firms in charge of pulling the strings in these sprawling webs of value creation, innovation is less about inventing and building physical things and more about orchestrating or coordinating good ideas.

Boeing and BMW are not giving up on innovation by any stretch. In fact, both companies are taking advantage of the resources they have freed up to focus on improving a few dimensions of value that matter most to their customers. More and more, these companies are focused on a new challenge as well: managing an increasingly seamless and supple fusion of design and development expertise from multiple suppliers, partners, and customers in global design and process collaborations.

We'll come back to these stories later. For a real eye-opening peak at the future of the global plant floor, let's take a ride on a peer-produced motorcycle. Our journey will take us to China, home to the world's largest and fastest-growing motorcycle industry. A signpost for the future of collaborative manufacturing, this industry is about as close to Linux as manufacturing gets (for now). In fact, you will be hard-pressed to find a recognizable company. Instead, motorcycles are manufactured by a self-organizing network of designers and assemblers that swap design ideas in the teahouses of Chongqing.

THE MODULAR MOTORCYCLE GANG

You might not have heard of it, but Chongqing is the world's fastest-growing metropolis, a rising economic hub in China and home to 31 million people. Situated near the Yangtze River, this former trading center now figures prominently in the government's plans to revitalize western China. In a single day, builders will lay 137,000 square meters of new floor space for residential blocks, shopping centers, and factories; over 1,370 people will take up residence in the swelling urban chaos; and the local economy will grow by ¥99 million (in yuan, or $12 million).

Somewhere in the midst of a dense haze of smog (Chongqing has one of the world's worst air quality problems), you will find Yin Mingshan, a sixty-eight-year-old industrial pioneer and a key figure driving Chongqing's relentless intake of people, money, and construction materials. Fourteen years ago he set up a motorcycle repair shop with nine staff members. Today his company Lifan employs 9,000 workers and has a turnover of ¥7.3 billion (over $900 million). Of course, he no longer repairs motorcycles; he makes over 700,000 of them a year for customers in 112 countries.

Apart from Chongqing, Lifan has plants in Vietnam, Thailand, and Bulgaria, and with distribution centers around the world, including one in the United States, its reach is increasingly global. Yin even plans to open a research center in Britain (where his daughter studies at Oxford). If all goes according to plan, Lifan will more than double its workforce to twenty thousand people within five years, and churn out enough motorcycles to make it a recognizable worldwide brand in the industry.

Not content to stop there, Yin is maneuvering to build up China's automobile industry. He's already made quite a name for himself, having purchased a BMW-Chrysler factory in Brazil. Now he's arranged to break the factory down, ship it up the Yangtze, and rebuild it in Chongqing.

Lifan already sells mid-size sedans in Asia, the Middle East, and the Caribbean. The Lifan 520 comes equipped with leather seats, dual air bags, a huge trunk, and a DVD system with a video screen facing the front passenger, all for $9,700. The next stop is Europe and North America. But don't count on world domination of the automobile industry just yet.

When it comes to motorcycles, Lifan is just one of many companies that have helped make China the king of the industry. Though you may not have heard of many of Lifan's compatriots, companies like Zongshen, Longxin, Jialing, Jianshe, and Dachangjiang are sharing in a remarkable success story that has seen motorcycle production triple from 5 million to more than 15 million vehicles a year since the mid-1990s. That's about 50 percent of the global pie, making China the world's leader.

Numbers only tell half the story. The characteristics that make the Chinese motorcycle industry successful also make for a particularly fascinating tale of how peer collaboration and production can give rise to potent competitive advantages, even in manufacturing industries where precision, efficiency, and quality control are mission critical.

Current thinking says peer production is only suited to creating information-based goods—those made of bits, inexpensive to produce, and easily subdivided into small tasks and components. Software and online encyclopedias have this property. Each has small discrete tasks that participants can fulfill with very little hierarchical direction, and both can be created with little more than a networked computer.

While it's true that peer production is naturally suited to bit products, it's also true that many of the attributes and advantages of peer production can be replicated for products made of atoms. If physical products are designed to be modular—i.e., they consist of many interchangeable parts that can be readily swapped in or out without hampering the performance of the overall product—then, theoretically at least, large numbers of lightly coordinated suppliers can engage in designing and building components for the product, much like thousands of Wikipedians add to and modify Wikipedia's entries. Sounds like a stretch, but the Chinese motorcycle industry provides a great illustration of how this works in practice.

Unlike traditional manufacturing industries, where tightly regimented production networks spit out end products under the command of a single leader, the Chinese motorcycle industry consists of hundreds of different companies that collaborate on motorcycle design and manufacturing. With only a modicum of hierarchical direction, these firms design and build new motorcycles faster and much less expensively than any conventional supply chain. Sounds like a recipe for chaos, but the industry has developed highly collaborative processes that enable local industry clusters in places like Chongqing and Zhejiang province to outdo their much more experienced rivals. The approach has been so successful that Honda, Suzuki, and Yamaha, once dominant in the region, have lost 40 percent of their market share to Chinese firms in less than ten years.

The story is more remarkable for the fact that just twenty years ago China had little if any domestic expertise to produce high-end motorcycles. Indeed, from the early 1950s, Chinese motorcycles were geared to rugged military needs. Assemblers and suppliers were state-owned and tightly regulated. Things stayed this way until the 1980s, when iconic Japanese firms such as Honda, Yamaha, and Suzuki were first allowed to enter China's burgeoning market. However, permission to enter came with a price. Japanese firms were not permitted to open up their own manufacturing sites; they could only license their technology to local manufacturers that were owned by the Chinese state. At first the Japanese had little reason to regret this, and they took advantage of China's bargain-basement labor costs. Their motorcycles quickly dominated the market, as their superior designs and better-performing products were a hit with local and regional consumers. By 1993, the state-owned manufacturers and their Japanese collaborators made China the largest producer of motorcycles in the world.

Then China's transition from communist central planning to a quasi-market economy added an unexpected twist to the story. Since motorcycles are considered less crucial to national development, the government experimented with a laissez-faire approach. When regulations loosened in the 1990s, private firms entered the industry at a furious pace, and soon overtook the state-owned manufacturers.

Whereas state-owned manufacturers cooperated cordially with the Japanese, the new breed of private firms were not so inclined. Chinese motorcycle makers had spent years mastering Japanese technology. Many Chinese workers learned about just-in-time manufacturing techniques, which equipped a large pool of local talent with the skills to operate high-end production facilities. Lifan is one of many companies that took advantage of the situation to grow rapidly, from tiny repair shops into full-blown motorcycle assemblers with products of their own.

Japanese firms question how much innovation is actually going on in Chongqing. They argue that Chinese firms simply knock off their products, and to a large degree they're right. Many of the most successful Chinese motorcycles are reverse-engineered Japanese designs, although the likes of Honda should hardly find this surprising. The Japanese have a long history of reverse engineering—just look at what postwar Japan accomplished when it reverse-engineered American automobiles and electronics.

So design innovations may not make the Chinese industry unique, but its process for undertaking a Japanese knockoff surely does. Normally, reverse engineering aims to re-create key elements of the original product as accurately as possible. A single firm chooses the product to be copied and issues detailed blueprints and goals to suppliers. In terms of organizational decision making, reverse engineering is usually no different from the approach most firms take to engineer a product from scratch. A lead product assembler makes most key design decisions and farms out the implementation to suppliers.

In Chongqing, several innovations have made the reverse-engineering process more collaborative and self-organizing. Historically, motorcycles have been high-performance products with integrated product architectures where each part is optimized to work in concert with the rest. For basic transportation markets in Asia, this performance orientation is overkill. The Chinese approach emphasizes a modular motorcycle architecture that enables suppliers to attach component subsystems (like a braking system) to standardized interfaces. This way, high-level designs are set out in rough blueprints that enable suppliers to make changes to components without modifying the overall architecture.

Rather than copy Japanese models precisely, suppliers take advantage of the loosely defined specifications to amend and improve the performance of their components, often in collaboration with other suppliers. Frame and fairing manufacturers, for example, work together to try out new designs in rapid succession before settling on one that meets overall cost, quality, performance, and integration objectives. At every step suppliers of adjacent parts take joint responsibility for ensuring that their components are compatible. Though some sourcing and marketing activities are global, the dense concentration of similar specialist shops facilitates the exchange of knowledge and skills among firms. Opportunities for face-to-face interaction engender additional trust, which is useful when production problems arise.

Together these three features play out in a self-organized system of design and production that Tokyo University economists Ge Dongsheng and Takahiro Fujimoto refer to as "localized modularization."[5] Design coordination is local as well as modularized in that suppliers of closely related components are responsible for delivering completed subassemblies. The process relies on the ability of suppliers to quickly test, develop, and retest how well their parts integrate with those of other local suppliers. But in the end it delivers a functionally equivalent motorcycle for less cost and in less time than the traditional top-down approach.

An outsider could be forgiven for assuming that such a decentralized system would wind up being incredibly chaotic and inefficient. But the experience of the Chinese motorcycle industry suggests the opposite. Modular architectures create the opportunity for increased specialization. Increased specialization in creating components drives innovation and improvements in quality and performance. Tight competition among hundreds of specialized suppliers keeps costs to a minimum. And close coordination among adjacent suppliers ensures that rapid development cycles deliver robust subassemblies to the final assemblers in tight intervals. Together, reduced time to market, low costs, and high-quality products have enabled the private assemblers to outperform their Japanese counterparts.

Highly collaborative approaches to manufacturing are not without risk. First there is a risk that suppliers and assemblers will gauge the market differently and create gaps in supply and demand. But with a variety of suppliers to choose from, assemblers can rely on multiple sources for common components, reducing the risks of undercapacity. Second, the lack of integration between suppliers and assemblers may lead to mismatched parts or suboptimal construction. Face-to-face relationships appear to be crucial to overcoming these challenges. In Chongqing, as in many other industrial clusters, informal networks share information about trends and market intelligence, and establish trust among a broad collection of employees and firms. People go to teahouses in their spare time, where entrepreneurs coordinate ideas for future design and knockoff projects.

Some say the proof is in the pudding, and by that measure China's motorcycle industry is a winner. Chinese manufacturers sold 10 million units in 1997, 11.5 million in 2001, and 15 million in 2004. The export of motorcycles reached 7 million units per year by 2005 (up from less than 500,000 in 2000) as the Chinese began threatening the Asian export markets that Honda, Yamaha, and Suzuki once dominated comfortably.

Today, China's motorcycle industry produces vehicles for India, Pakistan, Indonesia, and Vietnam, and nearly dominates the whole of the Asian market. Honda saw its share of the substantial Vietnamese market fall from 90 percent to 30 percent following the entry of Chinese motorcycle manufacturers. Chongqing assemblers offer functional technology for unbeatable prices—a value proposition that few in the Asian market will quibble with. In fact, over a ten-year period, the price of motorcycles built in Chongqing for Asian export has dropped from seven hundred dollars to two hundred dollars.

As the industry evolves, Chinese manufacturers are establishing brands, sales, and service networks to differentiate and gain higher profits. Economic forces will no doubt force industry shakeouts and challenge the localized modularization model. But even if the industry maturation leads to consolidation and some vertical integration, the cost and speed advantages of collaborative design and development will hold considerable appeal to future competitors.

THE LEGO BLOCK AIRPLANE

Phil Condit, former CEO of Boeing, used to explain how modern aircraft require constant replacement of various parts to keep them operational, saying that "a Boeing 757 is basically a bunch of parts flying together in close formation." For the next generation of aircraft his words ring even truer. These planes are basically a bunch of Lego parts from the get-go—provided by hundreds of different companies and assembled on a global plant floor in a giant, mind-boggling collaboration.

Here's what's happening. Innovation in the aerospace and defense (A&D) industry is about as complex and expensive as it gets. Like their cohorts in other R&D-intensive industries, A&D companies are finding that they simply can't access or own all of the knowledge necessary to compete. In fact, a global design and development team is increasingly a baseline requirement, just to stay in business.

At the same time, A&D companies need to be relentless in their bid to tackle ever-more complex engineering challenges without increasing costs. Airlines are perennially losing money, so anything that aircraft manufacturers can do to trim costs makes them more attractive partners for the airlines. Leading-edge A&D companies are responding with global, flexible, agile operational structures that enable continuous innovation, greater efficiency, and lower costs.

Some companies think and act globally by playing the mergers and acquisitions (M&A) game—they buy up companies with the capabilities they require and manage much of their innovation internally. Even the best-laid M&A plans, however, come with well-known integration problems and considerable costs to day-to-day operations.

Other companies, like Boeing, are moving in the opposite direction: shedding noncore assets and choosing instead to collaborate across global and loosely coupled value webs. Rather than the old hierarchical producer-supplier relationship, lead companies (prime systems integrators, in technical jargon) and their partners share the costs and risks of large development projects across the life cycle of new products and collaborate on everything from design to manufacturing, and even to long-term maintenance and support. The collaborative approach allows companies to tap best-in-class capabilities without the headaches accompanying the need to manage a full-blown merger or acquisition. Lead companies engage in less and less manufacturing and concentrate instead on designing systems and processes and orchestrating collaboration.

For Boeing, the change is part of a long, gut-wrenching process of transforming itself into a leaner and more focused competitor. Faced with a double whammy of the increasing instability in the post-9/11 commercial aircraft industry and a sharp decline in sales and market share, Boeing has been forced to revise how it does business. It's embracing mass collaboration—handing suppliers control over a large proportion of the thousands upon thousands of features and components that make up its airplanes in a bid to control costs, improve innovation, and get new planes to market more quickly. Partners participate in the design and building of the aircraft in ways not dissimilar to the programmers of the Linux operating system. While relinquishing a significant chunk of what was once the company's core manufacturing competency, Boeing is establishing a new competency in managing a globally extended partner base. In doing so, Boeing is using state-of-the-art collaboration technologies to transform a traditional collection of suppliers into a seamless global collaborative. The result is the development of the groundbreaking 787 Dreamliner, whose early sales and cost-efficiencies foreshadow a bright future for a once struggling giant.

While the sleek, fuel-efficient 787 is an amazing showcase of new technologies, the real story lies in how the plane has been created. Developing the 787 has taken Boeing's role as a systems integrator to the next level: building a next-generation aircraft with over one hundred suppliers across six different countries, in a truly collaborative fashion.

This is not simple outsourcing—Boeing has done plenty of that in the past. This time around, Boeing has built a broad horizontal network of partners who are collaborating in real-time, sharing risk and knowledge to achieve a higher level of performance. Like many of the companies we have discussed in previous chapters, opening up is about harvesting the best ideas and the best capabilities throughout the industry. This is a massive transformation for a company accustomed to intensely secretive and hierarchical ways of working.

In the past, Boeing's partners and suppliers didn't join the development team until the last stage of a detailed design phase. Boeing would design the specs and the supplier's job was to implement them. Everything was shipped to Boeing's plant in Washington. If parts didn't fit together, they had to be rebuilt. By the time the assembly process was complete, successive iterations of its planes had been built and refined on-site, with teams from around the world.

Boeing's new model treats suppliers as true partners and even peers, bringing them in much earlier in the process. In fact, before the 787 program was even announced Boeing was assembling an international team of aerospace companies to draw up plans for the new airplane. "We had over a thousand of our partners' engineering personnel here to jointly define the airplane," said Mike Bair, who heads up the 787 program for Boeing. "That way we get the best ideas from everybody, as opposed to just ours."

Deepening supplier involvement has significantly boosted the efficiency of the design process. Bair explains that when Boeing sent the specifications to the electronics supplier for the 777 (the predecessor to the 787) the document was twenty-five hundred pages long. "There wasn't a lot left to their imagination," he said. "We told them exactly what we wanted in excruciating detail." The equivalent specification document for the 787 is a mere twenty pages.

"We've realized that it's more effective when the people who are building the parts also do the engineering," says Bair. "They know better than us how their factories run, and to think that we can design a part that not only serves our needs but is also the most efficient for them to produce would be pure guesswork on our part."

The engine, for example, is being developed in partnership with General Electric and Rolls-Royce. More than twenty international systems suppliers (ranging from big names such as BAE in the UK to Matsushita in Japan to Honeywell, Rockwell Collins, and General Dynamics in the United States) will work with the Boeing team to develop technologies and design concepts for many of the various systems and subassemblies. As the design and development work concludes, the same companies will compete to become ongoing suppliers to the program.

Even potential passengers were invited to join the worldwide design team. When Boeing launched a Web site to promote the 787 it included a facility for aviation enthusiasts and other interested participants to describe the features they would like to see in an ideal plane.

Manufacturing is equally collaborative. When Boeing built the 777, it gathered together all ten thousand components at the project's end and assembled the plane at its plant in Everett, Washington. This time Boeing employees will be snapping together large components and subassemblies like Lego blocks rather than riveting and welding an entire aluminum plane. The modular approach will allow Boeing to cut the final assembly process down from the thirteen to seventeen days required for the 777 to as few as three days for the 787.

How do you build an airplane in three days? Many of the subassemblies—in fact, a full 70 to 80 percent of the new plane—will be wholly designed and manufactured by partners hailing from various parts of the globe. The vertical fin, for example, will come from Boeing's Frederickson, Washington, facilities; the fixed and movable leading edges of the wings from Tulsa, Oklahoma; the flight deck and forward fuselage section from Wichita, Kansas; the movable trailing edges from facilities in Australia; and the wing-to-body fairing from Winnipeg, Canada.

Japanese partners, including Fuji, Kawasaki, and Mitsubishi, are taking on 35 percent of the overall 787 structure, focusing on the wings and center fuselage. Vought Aircraft Industries of Dallas and Alenia Aeronautica of Italy are also involved, after having formed a joint venture to supply the tail section.

Altogether, it's a massive technological and human challenge to bring together such a diverse and globally distributed team of designers and manufacturers into a highly complex and structured development project. Underlying this complex network is a real-time collaboration system created by Boeing and Dassault Systemes called the Global Collaborative Environment. This cutting-edge system links all of the various partners to a platform of product life-cycle management tools and a shared pool of design data.

No more need to send engineering drawings back and forth between engineering and design teams. Any member of any team, anywhere in the world, at any time, can access, review, and revise the same drawings and simulations while the software tracks the revisions. Nonengineering managers can get in on the action too. Lightweight viewers enable everyone from marketing execs to cost accountants to review and comment on the plans as they progress, ensuring that the final design comes to fruition in the broadest possible context.

With more data from suppliers and more sophisticated software tools on hand, the virtual design process itself has become much more sophisticated than the tools used to design the 777. As Marcelo Lemos, president of Dassault Systemes (Boeing's software partner) explains, "We're moving beyond the digital modeling of static parts and geometry to the mechanical behavior of the plane throughout its life cycle, including its operations and maintenance."[6]

At the design stage, this level of sophistication enables various participants in Boeing's ecosystem to test how their components will work together in real-time simulations. Problems and incompatibilities can be identified long before anyone gets to the manufacturing stage. This in turn means that components that used to be designed serially can now be designed concurrently. Collaboration and concurrent design together are driving huge chunks of time and significant costs out of the system.

Take the 787 wing box, which Boeing is cocreating with Japan's Mitsubishi Heavy Industries. The wings for the 787 consist entirely of composite materials, which is a first for an industry that has traditionally relied on aluminum. Lightweight composite materials will contribute significantly to the 787's fuel efficiency, but they raised significant engineering and integration challenges for Boeing and Mitsubishi engineers. The team had to develop new engineering tools and processes and new manufacturing technologies and tooling in order to produce the new materials economically. Normally, a serial engineering process would take about six months. But all of these processes were done in parallel, using the new digital modeling tools, in just over six weeks.

The Boeing 787 will also include a health-monitoring system that will allow the airplane to self-monitor, alert onboard crew to potential problems in real-time, and report maintenance requirements to ground-based computer systems. If, for example, there is a problem with the wing, intelligent systems will detect abnormal vibration patterns and alert onboard crew and ground staff to take action.

When such problems arose in the past, pilots had to land the plane at the first opportunity, and call in an engineer to do a visual inspection and make a decision about whether to proceed with the flight or send in a repair team. Now engineers on the ground can make a remote diagnosis based on information beamed down by satellite. Ground teams can be mobilized and parts ordered well before the plane touches down. This saves precious time, which in the airline business means saving money. Boeing estimates that remote health monitoring will reduce maintenance costs by 30 percent.

More than the technology issues, it could be the intellectual property and knowledge-management needs that present some of the toughest hurdles to effective collaboration. "This project needs collaboration at the deepest level, and for that to be successful," says Lemos of Dassault, "we have to find the right mix of how much knowledge gets segregated as proprietary and how much will be shared."[7]

Most companies are understandably edgy when it comes to protecting their proprietary designs and processes. But in this project, sharing enough of the right information about designs and methods will be the difference between success and failure. "It's womb-to-tomb management of data associated with the airplane," says Bair. "Holding data back and not being forthright with where you're at and what's going on is unacceptable behavior. Everything is in the open here. We share everything," he adds.

Some engineers inside the organization worry that Boeing's extensive partnering and sharing of data may cause it to lose its engineering edge. There is a risk that too much know-how will leak out to partners and/or give rise to a powerful new competitor. The Japanese aerospace industry, for example, has long sought the knowledge to manufacture its own planes. Previous contracts have supplied most of the know-how, but companies such as Mitsubishi and Kawasaki still lack the technical prowess to develop the wings. Through the collaboration with Boeing, the Japanese may obtain their missing ingredient.

Though leaky knowledge is a risk in any partnership, it's a trade-off that companies can manage in order to reap the efficiency benefits of specialization and collaboration. "We keep a little bit of everything just for expertise," Bair says. Boeing, for example, chose to keep the design and construction of the vertical fin in-house. "We end up with a smaller, more capable, and more stable workforce," says Bair. "And then we look to other people who have better capabilities for those detailed parts of the airplane."

Managing these issues is all part of Boeing's new role as the leader of a global plant floor. "You have to be able to understand the market, you have to be able to turn that into requirements, you have to be able to integrate the partners and the parts to satisfy those requirements, and then you have to be able to support it in service," says Bair. "The knowledge that we have gained in organizing this program is a unique competency. I don't think there's anybody else that could do this, and we're going to do it again sometime on another new airplane program, so we'll be that much better at it."

The culture change from being an airplane manufacturer to a systems integrator has not always been easy. "There's a real danger that we're going to micromanage it," says Bair. "Once we get our partners to the point where we have a plan that will work, then the real challenge here is backing off, letting everybody do their jobs without us doing it for them." And yet, combining talents from around the world, while challenging at times, is a clear source of strength.

Block by block, the 787 project is already succeeding on a number of fronts. Most of the design work is complete, and the manufacturing is underway. The good news for Boeing is that airlines have finally embraced the 787 after a slow start in sales. In 2005 Boeing landed 354 orders worth more than $46 billion, beating Airbus in new airplane orders for the first time since 2000. The ultimate reward for Boeing, however, lies in validation of a new business model that is built around global collaboration. The 787 was a gamble and the company has certainly hit some glitches along the way with so many partners sharing so much responsibility. The company banked its future on peer collaboration, and given the sales of its new 787 Dreamliner, that gamble is paying off.

THE FABLESS CAR COMPANY

Automakers make cars, right? Wrong. Increasingly, premium companies design software, polish their brands, and piece together the complex electronic gadgetry that runs today's high-performance cars. In fact, next time you see a shiny new BMW X3 or 7 series drive off the dealership floor, you can be pretty sure that close to 70 percent of it was designed, built, and assembled not by BMW, but by a worldwide network of suppliers. Typical photos or TV clips of car companies these days show workers on the assembly line using precision robots to build the car. A picture of the X3 assembly would have to be taken at Magna International—the company that does final assembly of the vehicle.

This is a massive change. Not long ago BMW spent most of its R&D dollars on improving the mechanical infrastructure of its cars, like the drive train and the chassis. Such investments paid off, earning BMW a reputation for building well-engineered, performance-driven cars for the high-end market. These days innovation is shifting from engineering to a new digital frontier. While not giving up on mechanical engineering completely, BMW finds itself spending a growing proportion of its R&D budget perfecting the driving experience, and especially the software, electrical gadgets, and interfaces that drivers interact with.

BMW estimates 90 percent of its new innovations will come out of the fields of electronics and software. It's little wonder the company increasingly thinks of itself as an automotive brand wrapped around a set of design and software development capabilities. As BMW's head of development, Burkhard Goeschel, put it, "It would be fatal for us not to consider these areas of development as core competencies of BMW."[8]

This trend is mirrored around the industry. Mercer Management Consulting estimates that among the premium brands, electrical systems and electronics already account for more than half the vehicle's value. What's more, by 2015, suppliers, not the automakers, will conduct most of the R&D and production. Automakers like BMW will restrict their investments to a limited subset of components that are crucial to the success of their brands. This means more focus on the conceptual and design stages, and then on the customer experience and related downstream services. Everything in the middle will be outsourced or managed through one form of collaboration or another. Welcome to the global plant floor for automobiles.

This changing market reality is already deeply manifest in BMW's approach to innovation and operations. Many noncore design and manufacturing competencies are being hived off to partners and suppliers that manage everything from components to the final assembly. Goeschel says the company's intelligent outsourcing and cooperation model "frees up the financial and human resources to set benchmarks in the decisive brand-influencing fields of innovation."[9]

And yet, for a mid-sized automotive company intent on downsizing its share of R&D, it has impressive resources at its disposal. BMW has more than eighty-five hundred people in its global R&D network, not including suppliers, universities, research institutes, and increasingly, its own customers. The network reaches from Palo Alto to Japan, where specialized R&D units participate at various stages of innovation and production. The whole thing operates like a virtual conveyer belt that transfers know-how and skills across BMW's global workforce in order to boost agility in innovation and manufacturing.

In Japan, for example, BMW researchers work with local firms and universities to monitor, design, and test new automotive technologies ranging from engine components to electronic circuits. In Landshut, Germany, researchers specialize in lightweight construction technologies, and are always testing new materials and manufacturing processes. BMW employees in Palo Alto collaborate with local software houses and leading thinkers at Stanford and Berkeley to design next-generation automotive software. Back at BMW headquarters, "innovation councils" consisting of representatives from development, production, purchasing, and marketing are responsible for passing ultimate judgment about the potential for each new innovation.

Like other automakers, BMW has been eager to harvest the latent potential for innovation within its supplier community. Goeschel enthusiastically describes this potential as "limitless." A lot of the collaboration centers on the need to adapt the growing number of commodity components to meet the specs of BMW's premium brands. This means involving suppliers at an early stage, so valued innovations can be integrated into the design before BMW has fixed on a vehicle concept and architecture.

The front steering system in BMW's 5 series, for example, was co-developed with top-tier supplier Friedrichshafen. "Instead of requesting a completely mature concept from our supplier before we started the 5-series development phase," says Goeschel, "we brought the new technology to maturity jointly, and in close teamwork throughout the development."[10] Friedrichshafen worked on the hardware and basic software components of the steering system, while BMW worked on perfecting the software features that were relevant for customers.

In another collaboration, BMW is sharing the costs and risks of developing a new family of small gasoline engines with French automaker Peugeot. BMW's R&D department is in charge of designing the engines, while Peugeot is managing process development, production engineering, and procurement.

Goeschel claims that pooling know-how and leveraging scale economies in manufacturing will enable the partners to set new performance benchmarks without increasing costs. Up to one million small and mid-range Peugeot and Citroën vehicles, as well as future MINI Cooper models, will be fitted with the new engine every year. Bringing partners and suppliers into its innovation web enables BMW to harness innovation more quickly, and to continuously differentiate itself from its competitors. "It gives us the opportunity to rapidly integrate [their ideas] into vehicle concepts and bring them onto the streets," says Goeschel.[11]

Observers familiar with the industry will note that the rising importance of cooperation with suppliers is by no means unique to BMW, or new to the automotive industry for that matter. Suppliers already develop and build 65 percent of the average vehicle. But this percentage will increase to as much as 80 percent in the coming decades, making suppliers the main engine of growth and jobs in the industry.[12]

What's more interesting are the changes in the nature of the working relationships. As collaboration increases, automakers and suppliers will be closer and more intertwined than ever before. Goeschel describes it as a shift from the provider-customer relationship to a genuine development partnership that generates a new and more efficient division of labor. It's also a good way to ensure BMW retains the upper hand. As Goeschel points out, collaborating closely with suppliers allows BMW to keep on top of the necessary know-how about its outsourced components and services.

All considered, BMW has taken some important steps to harness the emerging global plant floor. To demonstrate true breakthrough leadership, however, it needs to go much further. Consider just three simple examples of changes that could vastly improve the industry. Note that these apply to all manufacturers of physical goods, and not just BMW.

Customer cocreation is underexploited generally, but it is next to nonexistent when it comes to manufacturing-intensive industries. We've already seen how collaborative sourcing represents a vital way for manufacturers to strip costs out of the system by involving suppliers and other stakeholders early on in the design stages. But why not also include customers, where their input would ensure that products are well adapted to their needs—not as an afterthought, or as a routine customization exercise, but as genuine codesigners who help shape the product concept itself. In Chapter 5, we briefly described the role that customers played in cocreating BMW's telematic features for future cars. BMW needs to do more of this. Indeed, there is no reason why every major innovation initiative should not involve customers as the central change agents.

With its innovation focus shifting to software, BMW and other companies could exploit open source software. While not suggesting that they open the kimono completely, we hasten to imagine what could happen if, like Google, eBay, and Amazon, BMW was to open up the API to its vehicle's software. We're not suggesting messing with the drive train, steering, or other functions that affect safety. Rather, thousands of independent developers could create new applications for work, life, and entertainment—key functions of the vehicle of the future.

Indeed, if Amazon can entice thousands of developers to add new services and applications to its platform, why couldn't an equivalent number help cocreate the digital environment that runs in future BMW models? What's more, we can easily imagine an entirely new business ecosystem emerging from such activities that would drive new innovation, create jobs and wealth, and add tremendous value for customers and automakers alike.

We also think that automotive companies, and manufacturers generally, are at least a decade behind other firms in other industries when it comes to thinking about how to harness Web-enabled marketplaces and ideagoras as enablers of innovation. Intellectual property exchanges that facilitate transfers of technology among a closed community of firms are not sufficient. Manufacturing industries—which, after all, share many of the same fundamental problems—need to invent and implement new ways to share and exchange knowledge.

Where, for example, is the InnoCentive for automotive engineers? Why can't BMW and its suppliers post problems on the Internet for tens of thousands of uniquely qualified minds to answer? Why rely solely on bricks and mortar when the Web provides a low-cost collaborative work space? Developing new ways to harness talent outside their boundaries remains an important and largely unexplored frontier for the auto industry.

HARNESSING THE GLOBAL PLANT FLOOR

The rise of peer and collaborative processes for designing and building physical things is not unique to the Chinese motorcycle industry, or to BMW and Boeing. These processes are emerging in industries where intellectual property is widely dispersed and production capacity is fragmented among hundreds of specialized firms. Increasingly, lead producers in fields such as semiconductors, computers, clothing, and bicycles are responsible only for product concepts, final assembly, and marketing. They outsource manufacturing and many, if not all, aspects of component design. And they rely on a global plant floor to tap into dozens or even hundreds of firms to help assemble finished products.

By organizing into loosely coupled networks of firms who jointly design and develop products for customers, both the suppliers and global integrators win. By taking on a larger share of the design and development work, suppliers increase their share of the intellectual property and profits in the final product. Global integrators gain speed and agility and can focus on high-value-added activities. Overall, this approach enables risk sharing and allows the network to tap into diverse skills and resources. In their book *The Only Sustainable Edge*, consultants John Hagel and John Seely Brown call this "productive friction": the new learning that takes place as knowledge and tasks are shared across enterprise boundaries.[13]

As planetary ecosystems for designing and building physical things spread, all manufacturing firms need to learn from the examples set by Boeing, BMW, and the Chinese motorcycle industry. What are the lessons?

Focus on the critical value drivers
Heightened competition and the increasing pace of change means that today's differentiating competencies can become overnight commodities, putting the entire value of your business at stake. Take note of where future opportunities for value creation are shifting and ensure that your capabilities are evolving in that direction.

If you're in the automotive industry, you'll find that these value drivers have shifted from the mechanics of the automobile to the driver interface, including all of the software and gadgets that shape the driver's experience. If you're in the aerospace industry, you'll find that controlling costs and decreasing the time required to get large-scale projects to market is more important than owning all of the requisite capabilities and engineering knowledge that contribute to the end product. Regardless of the industry, however, a new golden rule is emerging: Always strive to be the best at what your customers value most and partner for everything else.

Add value through orchestration

Companies with the capability to orchestrate collaboration on a global scale are still very few in number. Like Boeing, many manufacturing companies find themselves burdened by the inertia of past legacies. The process of weaning themselves off rigid and outmoded ways of doing business will create uncertainty and, at times, require employees to make uncomfortable adjustments. Boeing engineers may worry that they are giving up critical engineering skills and managers may question their ability to manage relationships at a distance, across companies, geographies, and cultures.

Despite these hurdles, there will be handsome rewards for those who learn the subtle art of weaving together the skills and competencies of disparate players to create globally integrated ecosystems for designing and making physical things. As Mike Bair of Boeing put it, "We are generating the unique capability to manage this extended partner base . . . one that we would not have developed had we not gone through all this angst with our partners."

Instill rapid, iterative design processes

A wide range of partners, each of whom is motivated to solve problems related to their key area of expertise, can accomplish rapid design and testing. We have seen this speed in open source software communities, and increasingly we see it in physical products such as motorcycle, cars, and airplanes.

In this case, the Chinese motorcycle industry is exemplary. Private assemblers and suppliers worked by trial and error to create an efficient division of labor that could replicate the best Japanese designs quickly and cheaply. Decentralization led to rapid iterations, experimentation, and informal networking among adjacent suppliers, while the ability to settle on core modular architectures enabled assemblers to integrate components and subsystems into finished products without having to impose much direction.

Harness modular architectures

The best way to stimulate innovation and iterative design in manufacturing is to harness modular approaches. Rather than mandating how to produce products, firms can work to create standards and modular architectures that specify product interfaces and leave it up to suppliers to get the job done. This is equivalent to Amazon's decision to open up its software APIs so that partners can add value to its platform.

A shift to modular manufacturing approaches means abandoning the view that outsourcing is just a way to off-load costs. Outsourcing is increasingly a way to gain speed, innovation, and knowledge. Whether you're BMW, Boeing, or Lifan, great global companies will fashion their products in a way that exploits the latent knowledge capital that resides in their web of suppliers and partners. They harvest best-in-class capabilities and collaborate closely to transfer skills and know-how across boundaries.

Create a transparent and egalitarian ecosystem

In the past, relationships throughout the typical supply chain were opaque and combative. Companies told suppliers to cut prices or lose their business. Buyers and sellers used whatever privileged information they had to gain short-term price, timing, and quality advantages over one another. Today suppliers increasingly act as partners, not adversaries. Undue secrecy, win-lose negotiating, and an insistence on exclusivity become counterproductive as the value suppliers add to the business web increases. The ability to create end-to-end visibility across the supply chain, on the other hand, can reduce costs, improve performance, and speed the metabolism of partnerships.

Global collaborators like BMW and Boeing understand that sharing information through interenterprise systems builds trust and helps enable networks of partners and suppliers to act as a single entity. At Boeing, in-person meetings are supplemented with video conferencing and the fully digitized, collaborative work space that integrates all 787 program partners into a single, seamless community. All partners understand that success or failure affects everyone, so it is in every participant's best interests to share critical business information. Without real-time access to a shared pool of design tools and engineering data, Boeing's efforts to create effective interenterprise collaboration would be nowhere.

Share the costs and risks

Sharing the risk of large development projects with partners spreads the cost and ensures that everyone is properly motivated. Boeing got this right by asking suppliers to share the up-front financial burden of developing the 787 Dreamliner. In return for sharing risk, Boeing's partners share in the upside. Cost savings all around make sure that everyone makes money.

As companies share costs and risks they should be prepared to share decision making as well. In the past, Boeing gave orders like a drill sergeant, and suppliers complied. Rarely did it matter if the supplier had a better idea—Boeing wanted the components built exactly as specified. This time, Boeing has given all its major partners a vote on matters that affect them. Engineers from Japan, Italy, and elsewhere are stationed in Seattle and participate in top-level decision making. Others routinely hook up via teleconference from around the world. Boeing and its partners are reaping the benefits as they work together on solutions and adapt the plan as needed to realize unexpected efficiencies.

Finally, keep a keen futures watch

Use your imagination. How will global plant floors and techniques such as localized modularization play out in markets such as health care, diesel earth-moving equipment, or construction? Build scenarios for these developments, and deploy the new skills you're learning to transform the world of atoms.

9. THE WIKI WORKPLACE
Unleashing the Power of Us

When Robert Stephens graduated from the University of Minnesota with a degree in computer science in 1994 he wanted to start a business consultancy. The problem was that good consultants cost a lot of money to hire, and Stephens had little, so he went into computer repair instead.

Stephens recognized early in his venture that the do-it-yourself crowd is a dying breed. From coping with the annoyance of computer viruses and spyware to the trauma of setting up a home network, a growing number of consumers trade cash for peace of mind by having a technician do the job. Stephens's answer to this consumer desire was Geek Squad, a cheekily branded service company that helps consumers navigate the increasing complexity of electronic gadgetry.

From very humble origins, Geek Squad grew and grew. Then, in 2002, the company was acquired by consumer electronics giant Best Buy after nearly a decade of profitable operations. Stephens had 60 employees at the time, and was booking $3 million in annual revenue. Over the next three years Geek Squad grew to 12,000 service agents, and under Best Buy's umbrella the division clocked nearly $1 billion in services from over 700 locations across North America, and returned $280 million to Best Buy's bottom line.

At age thirty-seven, Stephens is now leading the division in a bold effort to move Best Buy away from products and into services. Best Buy anticipates high-double-digit Geek Squad revenue growth in 2007 as the service offering benefits from volume and scale. CEO Brad Anderson describes Stephens's impact as huge, and says, "Robert Stephens has been at the heart of our service culture that we're building across our company."

Stephens is also teaching the old guard a thing or two about how to use the new universe of collaborative technologies to get the most out of Best Buy's employees. Geek Squad employees use wikis, video games, and all kinds of unorthodox collaboration technologies to brainstorm new ideas, manage projects, swap service tips, and socialize with their peers. They even contribute to product innovation and marketing. And all this makes Geek Squad a great place to work, and contributes to its stellar service record. We'll come back to those stories shortly, but first let's flesh out our hypothesis.

Just as the new Web is revolutionizing media, culture, and the economy, it is reshaping organizations and workplaces in a profound way. Peer production and cocreation are not just happening in online communities and networks like MySpace, Linux, and Wikipedia. Increasingly employees are using blogs, wikis, and other new tools to collaborate and form ad hoc communities across departmental and organizational boundaries. Geek Squad is just one of many examples in this chapter that signal the rise of openness, peering, sharing, and acting globally as fixtures of the future workplace.

The result is a number of deep, long-term transformations in the culture, structure, process, and economics of work. We are shifting from closed and hierarchical workplaces with rigid employment relationships to increasingly self-organized, distributed, and collaborative human capital networks that draw knowledge and resources from inside and outside the firm.

For people who toil away at desk jobs today, this prediction might sound outlandish. But as we explained in Chapter 2, a generation of young people are entering the workplace with a radically different philosophy of work. As eighty million young people in the United States alone enter the workforce they will bring high-technology adoption, creativity, social connectivity, fun, and diversity to the companies they work for, and increasingly, to the companies they found themselves.

Robert Stephens's Geek Squad is a great example of how technology and demographics are coming together in a radical workplace meritocracy that is rewriting the rules of engagement at Best Buy and showing the world how the new wiki workplace can produce superior bottom-line results.

GEEKS, WIKIS, AND GLOBAL DOMINATION

If you haven't already had occasion to call on Geek Squad's services, picture a cross between *Ghostbusters, Men in Black,* and *Dragnet.* Geek Squad agents carry a special agent badge and dress in black pants, white shirts, a black clip-on tie, and white socks with polished black shoes. All agents get special titles like Mission Controllers, Special Agents, and Black Ops, for the serious tech guns. To top it off, Geek Squad agents arrive at customers' homes in black-and-white Volkswagen Beetles.[1]

Together with the outfits and Geekmobiles, the James Bond titles foster a sense of fun and irreverence in an otherwise banal job. "Employees strongly identify with the brand," says Stephens. "They go on four to five emergencies a day, they drive a Geekmobile, they carry a badge—but it would all be pathetic and sad if we weren't profitable and we weren't really good service providers," says Stephens. "Agents even go grocery shopping with their Geek Squad jackets, and people walk up to them and ask them questions."

Geek Squad has got a lot of things right. Its brand, its systems, its business model, and its relationship with Best Buy are all part of Stephens's not-too-secret plan for world domination of computer services.[2] Take the Best Buy acquisition, for example. The potential for synergies was obvious.[3] Most service offerings in the computer industry are in shambles. And most customers dread wasting their time getting bounced from call center to call center. If Best Buy could turn the tech service into a sleek, profitable, and growing line of business, it would not only delight Best Buy customers, who could then count on quick and reliable service, it would benefit shareholders with increased revenue and earnings growth. Certainly, it was all upside for Stephens, who stood to gain access to a natural launching pad for a national computer services offering. Indeed, the customer response has been so great that Geek Squad precincts have now been stationed in nearly every Best Buy location across the United States and Canada.

The real Geek Squad secret sauce, however, is the people and how they collaborate. "We attract and retain talent longer, better, and more efficiently than anybody else," he says. Part of Geek Squad's ability to retain talent has to do with the branding and the fun workplace ethos that

Stephens has cultivated. Part of it has to do with traditional things like "hiring the best people."[4]

But branding and smart hiring aside, Stephens has learned to engage his agents in a continuous process of innovation and improvement that keeps the agents motivated to perform at their highest level. In fact, Stephens is full of great stories about the myriad ways his agents are the driving force behind innovation in Best Buy's services, and how the agents are always surprising him. The story that tops our list is the one about how Geek Squad agents instinctively started using online multiplayer games to stay in touch as the organization grew from 60 to 12,000 employees in just three years.

The ironic part of it is that Stephens had spent considerable time and effort building an elaborate internal wiki for exactly that purpose—to help keep all of the agents in the loop and to gather their input into the business. But the wiki was slow to take off, and Stephens was perplexed. He was always harping on the agents to use the wiki to communicate, but, at first, few of the agents bothered. Geeks are supposed to love wikis, so what was the problem?

Then one day Stephens asked a deputy director of counterintelligence at corporate how things were going in the field. "I worry about those agents in Anchorage, Alaska," he said. "There's about twenty of them there, and I worry about them staying connected to the mission." The deputy director said to Stephens, "Oh, those Anchorage guys, I talk to them all the time."

Curious, Stephens prodded him to reveal more details. So the deputy director sheepishly told him that they all play Battlefield 2 online. "With each server you can have 128 people simultaneously fighting each other in a virtual environment," said the director. "We wear headsets and use Ventrilo software so that we can talk over the Internet while we are running around fighting." Stephens, who now joins in himself from time to time, says the agents taunt each other, saying, "Hey, I see you behind the wall. But then, you know, while we're running along with the squadron with our rifles in our hands, one of the agents behind me will be like, 'Yeah, we just hit our revenue to budget,' and somebody else will be like, 'Hey, how do you reset the password on a Linksys router?' "

Stephens was aghast when he first learned of the agent's antics. "I just stood there in the hallway going, 'Oh my God,' I'm sitting here trying to build this shiny playground with all these tools for collaboration and I failed to notice what the agents were already doing. While I had my head down doing this in preparation to open the wiki's floodgates, the agents had self-organized online in probably the most effective and efficient collaborative tool that's already out there."

Stephens says that the agents now have up to 384 colleagues simultaneously playing at any one time. "They're talking and they're hanging out, and often they're talking shop and swapping tips," Stephens said. Geek Squad agents had just unofficially added another collaboration tool to the palette.

Stephens says the experience changed his thinking completely. "Instead of trying to set an agenda," he said, "I'm now going to try and discover their agenda, and serve it." Stephens even muses that he may get the agents to hack Battlefield 2 into a Geek Squad video game that he can use for training and recruitment.

By this time we were about halfway through our interview with Stephens. He had a lot more stories, each more compelling than the last. It turns out that for the Geek Squad agents, bottom-up communications was just the beginning. Next up was product development, as the agents applied their intimate knowledge of customers and technology to design award-winning products for Best Buy.

It all started when Best Buy decided to produce a new line of private label goods in China. They asked Stephens if he would allow them to put the Geek Squad logo on some of the devices. Stephens told Best Buy that he would do it, but only on one condition: Best Buy had to observe certain quality criteria, a bit like Martha Stewart or Ralph Lauren would. But above all, Stephens insisted that the Geek Squad agents should design them. No products would be released with Geek Squad logos without the agents' approval.

Best Buy agreed, but no doubt wondered if Stephens was crazy. He told the product developers not to bother hiring designers. "I want you to hire engineers who will execute the Geeks' actual designs," he said. Stephens rallied the agents to put sketches of potential products on the wiki. Hundreds of them did, and many more spent time providing feedback on the designs. "The agents love to debug, they love to criticize, taunt, and tear ideas apart," said Stephens.

Two months later the agents came back with a unique and pragmatic flash drive, which in today's electronics market is about as close to a commodity as you can get. And yet, the agents did something clever. Their design enabled the flash drive to fold into itself, so that it didn't require a cap. The agents knew that customers were always losing the cap, so a capless drive would be convenient. They also knew that nobody ever puts their flash drives on their key chains—not because they don't want to, but because the plastic rings are too thick and too hard. So the agents designed a flash drive with a thinly reinforced loop that slides easily onto a key ring. The design was good—so good, in fact, that in June 2006, Geek Squad won a prestigious German design award. "When the Germans give you an award for engineering, you know it's good," said Stephens.

Geek Squad agents have even come up with some of the company's most successful PR stunts. Stephens said that a few weeks before the new *Star Wars* movie was set to come out in the theaters, the agents were predicting that business would pick up. Why would business pick up, Stephens wondered? Because IT workers are the people you tend to see lining up at midnight to get advanced tickets. They end up staying up too late, and the next day they call in sick. Then, when problems occur in the workplace, their bosses call Geek Squad!

Stephens thought this was hilarious (and so did we). But the agents had more. They suggested the company fabricate an excuse note that IT workers could download up to a month ahead of time. They called the syndrome "prequelitis," and even established a trademark.

Next, Geek Squad sent out a press release predicting a mass wave of "prequelitis." IT workers and students, they claimed, would be reporting mass illness on March 28 (the day after the *Star Wars* movie was released). In the meantime, they posted the downloadable excuse notes on their Web site. They got over eight hundred thousand downloads, and a day later Stephens was on the *Today Show*. Stephens calls it "the cheapest money we never spent," and hails the fact that it was an idea that came straight from the Geeks.

"The PR value is nice," says Stephens, "but the real value lies in the sense of pride, identity, and purpose that is growing within the Geek Squad agents. There is a new, deeper level of self-awareness of their power as a group," he says. Stephens warns that group identity and purpose do not emerge overnight. "It took years to cultivate this culture. . . . It doesn't happen in a year."

Now that "agent culture" has taken root, Stephens says there is no telling where it will go. One thing is certain, however. When is comes to orchestrating employee collaboration, Stephens has a new rule: First observe, and then implement. "I'm deathly afraid of wasting time and energy trying to get people to do something they don't want to do. So next time, before I build that shiny, new playground, I'm going to think about how Geek Squad agents are already organizing—it's just much more efficient that way."

As for Geek Squad's prospects within Best Buy, Stephens is optimistic. "As long as there's innovation," he says, "there's going to be new kinds of consumer chaos."

RISE OF THE WIKI WORKPLACE

Geek Squad's successes signal the value of bringing high-technology adoption, creativity, social connectivity, fun, and diversity into the workplace. But are ideas behind the wiki workplace really new?

There has long been recognition that organizational bureaucracy impedes innovation, agility, and success. Walk into a typical office less than a century ago and one would expect to see long rows of desks, regimented in army fashion, with typists clicking away from nine to five—all under a managerial ethos that borrowed heavily from the military's command-and-control structure.

For half a century there have been successive theories and attempts to free the creativity of human capital. Most of these management theories were predicated on the view that computers could change the ways organizations work. In 1962, Douglas Engelbart wrote an extraordinary paper entitled "Augmenting Human Intellect: A Conceptual Framework," where he explained how electronic workstations could augment the thinking and communications abilities of what he called "knowledge workers." The theme of teamwork was big in the eighties, and empowerment and networking were big in the nineties.[5] But what has really changed?

The record shows that corporations have become networked in the sense that they build business webs with partners on a platform of information technology. While this is a huge development, fundamental changes in the internal structure and management of organizations have been illusive. Jeff Pfeffer of Stanford's Graduate School of Business says, "For fifty years various people have speculated about how the advent of computers was going to change the workplace—the distribution of information would delayer and decentralize organizations and management. Except for relatively few organizations this hasn't really occurred." He argued, "Traditional hierarchies still exist. Bosses still expect to be bosses. Command and control is alive and well." To Pfeffer, this partly explains why there is considerable dissatisfaction in the workplace today.

However, the new business environment, the Net Generation, and the rise of the new Web are finally beginning to change all this. Most large organizations today are geographically dispersed. This fuels a need for people to communicate and work together while being separated by great distances. Networking technologies allow companies to run cohesive yet decentralized operations by linking employees in virtual teams and communities of practice.

Competitive pressures, meanwhile, are making organizations leaner and more agile, more focused on the customer, and more attuned to dynamic competitive strategies. This means firms are less hierarchical in structure and decision-making authority than they used to be. But it also means that they will be less likely to provide lifelong careers and job security, and more in need of continuous reorganization to maintain or gain competitive advantage.

At the same time, the nature of work itself is changing. Work has become more cognitively complex, more team-based and collaborative, more dependent on social skills, more time pressured, more reliant on technological competence, more mobile, and less dependent on geography. Many employees are already given far more autonomy to decide how and where they want to work. A growing number of firms are decentralizing their decision-making functions, communicating in a peer-to-peer fashion, and embracing new technologies that empower employees to communicate easily and openly with people inside and outside the firm.

The continuous flow of new technologies into the workplace has been a key source of change in the way that we work. For members of generation X and earlier, the most definitive workplace changes began with the rapid convergence between office telephone systems and computer networks. E-mail enabled employees to share information far more efficiently than they could with typewritten memos. Client-server computer architectures gave them access to company data that used to be guarded jealously by senior managers. Cell phones and BlackBerrys gave staffs the ability to work on the move and spend more time out of the office.

Finally, today, a younger generation of workers is embracing new Web-based tools in a way that often confounds older generations but promises real advantages for companies that adapt their style of working. Tools such as blogs, wikis, chat rooms, peer-to-peer networks, and personal broadcasting are putting unprecedented power in the hands of individual workers to communicate and collaborate more productively. This in turn is driving a new revolution in workplace collaboration of a qualitatively different nature.

Having matured quickly in the last three years, these weapons of mass collaboration enable employees to engage and cocreate with more people, in more regions of the world, with a richer, more versatile capability set, and with less hassle and more enjoyment than any earlier generation of workplace technologies. Employees can act globally too—cutting across organizational silos and connecting with customers, partners, suppliers, and other participants that add value in the firm's ecosystem. What's more, the increasingly open source nature of these tools means that this new infrastructure for collaboration is accessible to a much wider base of people and businesses—so wide, in fact, that there are very few barriers to adoption for organizations of any persuasion.

To add further fuel to the fire, a new demographic is arriving in today's workplace that cannot imagine a world without Google or mobile phones. The Net Gen has experienced these inventions and breakthroughs as part of their birthright, unlike earlier generations who have had to adapt or acclimatize to instant messaging and the iPod. Having been nourished on instant messaging, chat groups, playlists, peer-to-peer file sharing, and online multiplayer video games, they will increasingly bring a new collaborative ethos into the workplace. Working together and sharing their knowledge across organizational boundaries—in much the same way as they swap songs and videos over the Internet—will be perfectly normal for tomorrow's workforce.[6]

Of course, the new workplace is about much more than wikis and other technologies, just as wikis are about much more than Wikipedia. The Net Gen also has a unique set of formative experiences that shape their sense of workplace norms and values. When asked about the experiences that define their worldview, members of the N-Gen talk about the fall of the Berlin Wall, the Rio Earth Summit, and 9/11. All of this adds up to a profoundly different attitude and approach to work and a unique set of expectations of their employers. Whereas previous generations value loyalty, seniority, security, and authority, the N-Gen's norms reflect a desire for creativity, social connectivity, fun, freedom, speed, and diversity in their workplaces. Attracting, engaging, and retaining these employees in an increasingly competitive environment will demand that companies understand the Net Generation and the individuals who will emerge as its leaders.

BEST BUY'S BOTTOM-UP INNOVATION

After talking to Robert Stephens and hearing the incredible story of the Geek Squad we then spoke with Best Buy CEO Brad Anderson. The $30-billion-a-year company has ruled consumer electronics retailing for about ten years now, and there are no signs of its category dominance abetting. So we wanted to see what was up. We learned that Geek Squad was only one of many tales of workplace innovation at Best Buy, and that there were more audacious leaders like Stephens making waves inside the company.

When we spoke, Anderson was in the midst of rolling out a bold new "customer centricity" strategy in a bid to boost profits. In a nutshell, Best Buy figures out which customers make it the most money, segments them carefully, then realigns its stores and empowers employees to target those favored shoppers with products and services that will encourage them to spend more. According to Anderson, the key to getting everything right was an enterprising group of sales associates and managers led by Gil Dennis—a bright and passionate store manager with a radical idea for his employer.

Dennis believed his peers across the company probably understood Best Buy's customers in ways head office didn't. But until Dennis came on the scene, Anderson's customer centricity strategy had relied heavily on market researchers who poured through reams of sales and demographic data to determine how to optimize Best Buy's store in each location. Store managers and associates were already getting to know customer habits, wants, and frustrations more intimately than can be conveyed in market research statistics. Their knowledge of the surrounding community could assist Best Buy in fine-tuning its customer centricity strategy to the needs of the local demographic.

Dennis himself had many ideas about how Best Buy might enhance its day-to-day operations. But how, thought Dennis, could these employee insights travel from the front lines to the core of Best Buy's strategy? In most companies Dennis's idea would have had to work its way up the bureaucracy before some minion with authority could consider it. The idea would be quashed 99.9 percent of the time before it got anywhere near a decision maker. Dennis wanted an open forum where all Best Buy store managers could communicate with each other and influence the CEO.

Dennis might have been politely thanked for his enthusiasm, encouraged to stay focused on his job, and told to climb the company ranks before his ideas would have merit. But Brad Anderson caught wind of the idea, and it triggered a revelation. His team wasn't doing enough to tap the vast and intimate knowledge of Best Buy's employees—especially those employees on the front lines who engage with customers on a daily basis. "We needed to use the entire range of human capacity within our business to generate those customer insights," said Anderson. "Getting frontline employees involved in customer centricity was crucial to both the design and execution of the centricity strategy." Part of the challenge in executing the strategy, however, was finding an efficient way to share, evaluate, and apply the insights that thousands and thousands of Best Buy employees were harboring about what customers need and how best to serve them. It seemed that Gil Dennis had the answer.

With support from Kal Patel, Best Buy's executive vice president of strategy, Dennis organized a meeting of some other general managers, and they called it the GM Forum (now the Retail Leadership Forum). Dennis facilitates the meetings, and along with other store managers has initiated a broader collaborative process that is generating some great insights. Dennis's initiative to engage the field has become a key part of the management process at Best Buy. Various corporate functions now gather the latest customer insights by talking to the Retail Leadership Forum on an ongoing basis.

"It seemed like a very natural thing to do," said Patel, "for senior executives to spend time with people who are interacting with customers every day, and to let that shape how they direct the company and how they determine strategy." And yet, Patel and other leaders at Best Buy discovered that cutting across the organizational hierarchy was quite unsettling at first. Patel observes, "When employees are living in a hierarchical structure there's a lot of fear. People two or three layers above you resist the rules being changed. And with all that fear most people do nothing. They let the hierarchy rule."

Management innovations like the Retail Leadership Forum change the rules with respect to whom people talk to, how they talk to each other, and what they talk about. In most workplaces talking to your boss's boss is a no-no. Disrespecting the established channels of authority might even get you fired.

Young employees like Dennis have less reverence for organizational protocol. Patel describes Dennis as "relatively young in his career, with a naive sense of idealism and a ton of raw energy for change. Dennis hasn't totally embraced the orthodoxy of business structures and all the things that beat people up inside large businesses."

Dennis's infectious enthusiasm for getting people to connect across the company has ignited a new way of executing the customer centricity strategy. Anderson sees the Retail Leadership Forum as a way to open up the company to new kinds of interaction and knowledge sharing. "We're drilling holes through the bureaucracy and giving people a safe place where they can create a community of shared knowledge," says Anderson. "We needed to make that knowledge available to people inside the system and to try to keep it as impervious to the influence of the bureaucracy as possible."

Now instead of dictating the customer centricity strategy from on high, Best Buy gives employees autonomy to develop their own strategies. It requires innovation at the store level and general managers fine-tuning the company's broad-brush thinking to meet the needs of local demographics. To a company that had relied on a classic merchandising model that flogs new products at customers with a cookie-cutter approach, this was a radical change.

Every associate is encouraged to try new ways to increase Best Buy's sales and profits. They are rewarded financially when they succeed, and in lots of other ways just for trying. Patel likens Best Buy's bottom-up approach to innovation to the scientific method: "Every one of our associates is trained in a method that starts with a hypothesis, and then proceeds to stages like test and verify. At the end of the experiment, the associate who undertook it reports out on what he or she has learned."

"These are quick 'popcorn-stand'-like experiments," says Patel. "If we want to try something new, we might pitch a tent in the parking lot of one of our stores and test an idea out with our customers."

Some business thinkers claim that the bottom-up approach to collaboration and innovation is often counterproductive. They warn that "letting a thousand flowers bloom" ends up giving companies a lot of weeds, diluting their focus on the few big ideas that are going to "move the needle" in multibillion-dollar companies.

But Anderson counters that without these systematic, cross-company forums much valuable knowledge inside the organization would go unutilized. "Getting other points of view and other pieces of knowledge into our learning system that might otherwise have escaped is key to our success as an organization," said Anderson.

At the same time, frontline employees get a chance to see how their views mesh with the perspectives of people inside the bureaucracy that "might have merit that store managers are not paying attention to," says Anderson. Store managers can then communicate those perspectives back to employees in a way that the bureaucracy couldn't.

SOCIAL COMPUTING IN THE ENTERPRISE

Best Buy demonstrates how drilling holes through the hierarchy of an organization can produce great results. But what happens when you mesh Brad Anderson's philosophy with a powerful new infrastructure for collaboration that includes wikis, blogs, and RSS? To find some answers we talked to Ross Mayfield, CEO and founder of Socialtext, one of a growing number of start-ups that have emerged to supply social computing technologies (especially wikis) to enterprises.

Socialtext itself is emblematic of the new organization and the new

approach to work. The entire company is virtual. They have ten employees, zero overhead, and everybody works out of their homes. Socialtext
doesn't spend a dime on traditional marketing. But they've managed to
cobble together funding from angel investors that has kept them on a
shoestring budget and hungry to drum up business.

"Every single thing we do is tapping into the network that we've been
able to create, and then also eating our own dog food using a combination
of Socialtext, Skype, and FreeConference to run our entire company," says
Mayfield. "We've developed a business model—through blogging, social
networking, PR, and through our product—where demand comes to us."
When we last spoke with Ross, he had over four hundred customers, of
which nearly thirty were Fortune 500 firms.

That demand, if anything, is a product of the buzz that has been generated around the new way of working and collaborating enabled by Socialtext's enterprise software. Mayfield has a bold vision of the future of
the organization, and it's largely his enthusiasm and tenacity in pursuing
this vision that is driving the company forward. The vision springs from
complete disillusionment with traditional workplace collaboration tools.

"For a long time," says Mayfield, "personal productivity tools and
applications—the kind that Microsoft makes—have been centered on a
single user who generates documents. You also have highly structured enterprise systems designed and implemented from the top down—in many
ways as an instrument of control—with rigid work flow, business rules, and
ontologies that users must fit themselves into. The problem is that users
don't like using those kinds of tools, and what they end up doing is trying
to circumvent them. That's why ninety percent of collaboration exists in
e-mails."

Mayfield argues that traditional organizations have reached a point
where e-mail itself is breaking. "You could argue that ten or twenty percent of e-mail is productive," says Mayfield. "Your average Fortune 1,000
employee spends four hours a day in their in-box, and we've got to find a
way to get them out." Anyone who has to deal with a daily flood of occu-
pational spam can surely sympathize with Mayfield's view.

Mayfield thinks the solution is collaboration tools that adapt to the
habits of workplace teams and social networks rather than the other way
around. Indeed, this insight was the genesis of the business. He and his

partners came up with the idea for Socialtext when they observed how employees in Silicon Valley firms were bringing in applications from the open source community and using them as a source of personal competitive advantage. If the tool proved to be effective, they would quickly begin to see a bottom-up demand pattern emerging in the firm as other employees clamored to try them out.

This "in-through-the-backdoor" approach to technology adoption is not particularly new in the workplace. It happened with e-mail, and especially with instant messaging—a technology that many organizations found threatening at first. Now e-mail and instant messaging are workplace standards, and the same thing is happening with wikis. According to Tim Bray at Sun Microsystems, this is a lesson we ought to have learned by now. "The technologies that come along and change the world are the simple, unplanned ones that emerge from the grassroots rather than the ones that come out of the corner offices of the corporate strategists," he says.

John Seely Brown, former chief scientist of Xerox and director of its PARC, says, "A lot of corporations are using wikis without top management even knowing it." "It's a bottom-up phenomenon," he adds. "The CIO may not get it, but the people actually doing the work see the need for them."[7] As a growing number of CIOs see the benefits, however, they are beginning to give workplace teams license to experiment with new social-computing technologies.

At Dresdner Kleinwort (DKW), a Europe-based investment bank, employees started using wikis in the IT department to document new software in an informal pilot. Soon afterward, wikis began to migrate out of the IT department and into the broader workplace environment, where teams picked up on them as a way to get collaborative projects up and running quickly.

When DKW CIO JP Rangaswami learned of the process, he was intrigued by the technology's versatility. The company went ahead with more pilots, and after just six months of usage, the traffic on the internal wiki exceeded that on the entire DKW intranet. Today the wiki has more than two thousand pages, and is used by more than a quarter of the company's workforce. Lead users have decreased e-mail volume by 75 percent and cut the company's meeting times in half. Rangaswami says, "We recognized early on that these tools would allow us to collaborate more effectively than existing technologies."

At Xerox, the company's chief technology officer, Sophie Vandebroek, is using a wiki to collaboratively define the company's technology strategy. Normally, high-level strategy documents are created in a hierarchical fashion, where the boss controls the vision and content. Vandebroek decided to turn everything inside out by opening the process up to all researchers in the R&D group. Vandebroek expects more robust technology road maps and a much stronger competitive strategy section as a result. "We'll get more content and knowledge in all of our areas of expertise," she says, "including everything from material science to the latest document services and solutions."[8]

Mayfield suggests that part of the reason wikis are popular and useful is inherent in the nature of the collaborative tools themselves. "They have very different properties, because they ask users to share control, and that actually fosters trust. The more participation that you have," he continues, "the greater quality you'll have in a project, in the same way that open source works."

Many wiki users and aficionados say the benefits are linked to the ease and efficiency with which collaboration takes place. Tantek Çelik, Technorati's chief technologist, uses wikis for everything—his work, his social life, his voluntary activities, and for staying in touch with family. He says wikis distribute the burden of organization across a collaborative network instead of making an individual project manager a choke point. "Now everyone can make incremental progress without having to wait for everyone else," he said. "It's like parallel processing for people rather than computers." Çelik goes so far as to suggest, "The ability to use wikis will be a required job skill in five years."

Everyone we spoke to about wikis in the course of our research agreed that the trust and efficiency benefits of social-computing technologies are evident. But most also agree that more profound opportunities lie in the ability of organizations to experiment with nontraditional workplace design philosophies. As Robert Stephens learned with Geek Squad, real innovation can occur when companies take the time to observe how the existing workplace culture operates in a "state of nature," and then learn how to serve that culture effectively. This means ending the practice of trying to force employees into rigidly structured work-flow tools that stifle their creativity and encumber them with complex processes and architectures.

In contrast to complex group collaboration tools, wikis conform naturally to the way people think and work, and have the flexibility to evolve in a self-organizing fashion as the needs and capabilities of the organization change. This flexibility arises from the fact that at their most basic, wikis are completely unstructured. "The structure," says Mayfield, "is created by demanding active involvement from users in ways of organizing and creating their own information architecture."

So rather than begin with a top-down ontology, process, or taxonomy, employees can fashion their own structure as required with a bottom-up, collaborative process. "Wikis hand control over to users to create their own ways of organizing knowledge, workplaces, processes, and perhaps even their own applications in ways that they've not been able to do before," says Mayfield.

The bottom-up approach that Mayfield and others advocate challenges the deeply engrained notion that employees are helpless without the aid of clearly and rigidly defined structures to guide them. The fact that such notions persist, however, highlights the degree to which we have failed to update our assumptions and biases as the nature of work changes. We still, for example, think in industrial-age terms about work as a routine that repeats endlessly. Even as work has become more cognitively complex, we still entertain visions of knowledge workers spending much of their time pushing paper.[9]

Yet the vast majority of employees don't do business processes anymore, at least not in the traditional sense. After years of optimizing supply chains, outsourcing, automation, and stripping costs and inefficiencies out of the back office, most employees spend very little of their day working on regularized activities. "What they do," says Mayfield, "is they manage exceptions to processes. Even in the most mundane workplaces like a call center, people are constantly wrestling with new problems."

When new problems and exceptions arise, people in organizations will swarm around that exception to try to resolve it. Think about the last time something in your workplace went haywire. How many people were jumping up to help solve the problem? In most workplaces, the answer is "as many as possible," because people genuinely enjoy the challenge of coming up with solutions to workplace exceptions in a truly spontaneous and collaborative fashion (it definitely beats the 9:00 A.M. meeting!).

The problem from an organizational and knowledge-management point of view, however, lies in the inability of firms to capture and codify those moments of inspired brilliance—the moments when someone does something spontaneous that could be the key to unlocking a whole new approach to getting things done. Mayfield suggests the self-organizing group formation process should occur in social software. "Those are the moments where the greatest amount of learning occurs," he says.

In a traditional workplace, this decentralized approach to problem solving might be worked out in the lunchroom, while leaning over a colleague's cubicle, over a pint after work, or increasingly through a long thread of e-mails. The problem is that this casual approach to problem solving leaves no organizational memory of the event, with the risk that only the people involved in creating the solution walk away with any new insights. Problems can persist like a bad cold, and solutions will be reinvented every time the problem reoccurs.

Social software provides companies with a way to document and leverage those moments of innovation with relative ease, providing a living, breathing repository of easily accessible knowledge that grows along with the organization. Companies can continually harness their local insights and adaptations to new problems by capturing and using those insights to drive organizational change and renewal.

"You release early and release often," says Mayfield, citing the open source dictum. "When you come across a bug in the workplace, you have an ethic of fixing it right then and there so you have these tight little iteration cycles. Wikis compel teams to engage in a constant state of rapid prototyping."

The world of software and Web services may have set the standard. But, in this case, we're not just talking about software. This iterative and collaborative approach to innovation is how the entire economy will run from this point forward: rapid incremental innovation, over and over and over again. Every product, project, or service is in perpetual beta mode—a state of continual refinement and improvement as employees, partners, and suppliers pool their knowledge and capabilities to meet the evolving needs of customers. "In the end, nothing is in an end state," says Mayfield. "Even with Wikipedia, the best thing you could say today is that it's better than it was yesterday, and tomorrow it's going to be even better. The project is never going to end."

PEER PRODUCTION IN THE WORKPLACE

We've now heard from Brad Anderson and Robert Stephens about Best Buy's bottom-up approach to innovation. Ross Mayfield from Socialtext told us about social computing and how simple technologies like wikis are giving rise to new forms of workplace collaboration. So far, so good. Now what if we took this a step further and embedded a full-fledged, no-holds-barred version of peer production in today's enterprise? How would work and life be different?

Our research revealed that this is an exercise in fact-finding not fiction. Companies that increasingly seek to leverage external knowledge and resources must ask new questions about how to manage their workforces. What kinds of talent and expertise should they seek to retain inside their boundaries, and what should they seek to harness externally? How do they weave external and internal resources together? And to what extent can the new models of peer collaboration and production discussed in this book serve as a model for managing these hybrid workforces?

Think back to some of the key examples in earlier chapters. Large companies like Amazon, Boeing, IBM, P&G, Merck, and others know that keeping up with customer demands for faster innovation and greater customization means sourcing a lot more innovation from outside their corporate walls. Some of this will come through proprietary channels and networks like licensing, outsourcing, and joint ventures. But a great deal more will come from more open and amorphous networks of peers.

Many of the people who participate in corporate ecosystems will not work for them. There will be no contractual relationship, so they can't directly control them, nor can they expect to own or monetize all of their intellectual property. IBM, for example, has no legal contracts with the Linux community. So instead of "managing" this extended development team in a traditional top-down fashion, IBM is learning a new participatory management model where many key decisions, resources, and activities are shared with the community.

In fact, the company took this new workplace philosophy a step further in September 2006 by inviting employees from more than 160 countries—along with their clients, business partners, and even family members—to join in a massive, wide-open brainstorming session it called the InnovationJam. Over the course of two seventy-two-hour sessions IBM engaged over one hundred thousand participants in a series of moderated online discussions. Their combined insights surfaced breakthrough innovations that IBMers expect will transform industries, improve human health, and help protect the environment over the course of the coming decades. CEO Sam Palmisano believes so strongly in the concept that he's committed up to $100 million to develop the ideas with the most social and economic potential.

With increasing access to external capabilities—and clever collaborative mechanisms like the InnovationJam—many mature organizations will opt to employ smaller and much more decentralized teams, whose principal role will be to orchestrate value creation rather than participate directly in it. Their job will be to identify and broker deals with the communities where exciting things are happening, wherever they happen to be in the world. These new workforces will need to craft the incentive systems that enable firms and their collaborators to reap a fair share of the value. They will also take responsibility for functions where hierarchical control of mission-critical apps is still important. You may not want to self-organize your inventory or accounting, for example.

Companies will need to identify the leaders within their organizations who are capable of orchestrating these amorphous networks of peers. Everyone else will plug-and-play, adding value and then moving on to other projects. The end result might look more like a Jackson Pollock painting than a traditional organizational flowchart, but this much more granular and collaborative division of labor will enable a more flexible and fluid approach to innovation and value creation. That in turn will make companies successful in a highly turbulent and competitive environment.

The bottom line is that the workplace is becoming a self-organizing entity where centralized and tightly controlled processes are increasingly giving way to more spontaneous and decentralized forms of mass collaboration. To see how this might pan out, we took a look at five typical workplace functions: teaming, time allocation, decision making, resource allocation, and communications.

Teams

In the old days you were assigned to a corporate team, and that's where you stayed, building up bonds of trust and loyalty that would enable you and your teammates to collaborate effectively. Today, new forms of mass collaboration suggest that companies may be better off with a more self-organized approach to teaming.

Over 16,000 people are actively peer-producing Wikipedia. A quarter of a million people collaborate on Slashdot. Thousands of programmers contribute to Linux. On Amazon, 140,000 developers are building applications and businesses. These large-scale efforts don't employ teams, at least not in the traditional sense. They "employ" peer-to-peer networks with a constantly changing roster of participants.

Could such a highly federated and highly fluid approach to teaming ever fly in a traditional workplace? Most firms assume not and opt for the U.S. military standard of 150 people as the ideal size for an operating unit.

And yet, an increasing number of employees work from home or on the road. Some companies have even done away with specific desks for employees, who set up their tools of the trade wherever they find a vacant space and clear away their things as soon as they leave. Some teams form temporarily around a particular project, in much the same way that film crews come together for a few months at a time to produce a movie, and then go their separate ways to work on other productions.

Some will no doubt complain that such a radical approach to workplace organization is unmanageable. But if that were really the case, then we wouldn't see communities like Wikipedia, the open source movement, or the Human Genome Project collaborating successfully on a very large scale. With the right tools and enough transparency, a large and diverse group of people self-selecting to add value can complete even the most complex tasks with only a minimum of central control.

Time Allocation

If you work at Google, what are you required to do with 20 percent of your time? Goof off! The company directs employees to dedicate 20 percent of their time to personal projects—projects that interest employees but needn't slot neatly into Google's predefined product road maps. In keeping with its belief in collaboration and encouraging self-organization, the company tracks the pet projects that employees conjure up.

Company officials reason that although Google employees are only a small fraction of the programming talent in the world, they are among the brightest programmers in the world. So in addition to leveraging the insights of external developers (as explained in Chapter 7), Google allows its employees to pursue their own interests. This not only makes them happy, it boosts creativity and can surface unplanned innovations that may one day evolve into successful business ventures.

Google CEO Eric Schmidt told us he hadn't had a product idea in years. "Virtually all of the product ideas in Google," he says, "come from the twenty percent of the time employees work on their own projects." One such innovation is Orkut, a social-networking service named after its inventor Orkut Büyükkökten, a Google software engineer who developed the project during his allotment of personal time.

Decision Making

Ronald Coase liked to describe firms as islands of hierarchy within a sea of decentralized market activity. Today it's like the invisible hand of the marketplace is being extended right down to the actual worker, rather than merely to the industrial sector or firm. Markets are being put to work in corporate strategy, planning, and execution to allow the insights of a much broader and representative group of company stakeholders to inform company decision making.[10]

Companies including Hewlett-Packard, Eli Lilly, Siemens, and Microsoft, for example, have used internal prediction markets to forecast product sales, identify promising drug candidates, and spot emerging trends and technologies. In essence, these companies ask a question, then invite as many people as they like—including employees, partners, suppliers, customers, and other knowledgeable participants—to buy and sell "virtual" stock based on their confidence in a particular outcome (i.e., this product will be a winner with soccer moms). The result is a trading price that tracks the consensus opinion over time, reflecting new information and changing circumstances on the ground. In each case, the predictions generated by internal markets have been as good or better than official company estimates.[11] For example, Kay-Yut Chen, principal scientist for HP Labs, says that prediction markets run to forecast annual computer workstation sales outperform internal corporate forecasts in six out of every eight cases.

The great thing about prediction markets is that they provide a relatively low-cost, self-organizing approach to tapping collective intelligence, both inside and outside the firm. But few companies would make markets their sole source of decision inputs. For one, there is the problem of the herd. "The wisdom of crowds" breaks down when they lose diversity or people stop thinking independently and start following the group. Yet, on average, well-structured prediction markets provide forecasts that outperform those of even the best-informed and highest-paid experts.

Resource Allocation

Similar market-based processes are being extended to tasks like resource allocation within firms. The idea is simple: Resources ranging from spending budgets to computing power are tradable commodities, so why not allocate them with a marketplace approach that ensures they go where they are most valued? It largely removes internal politics from the process and sets up a dynamic where teams buy and sell access to resources based on their independent judgments about how badly the resources are needed.

At Hewlett-Packard a group led by Bernardo Huberman at HP Labs is testing internal markets that let employees buy and sell rights to use shared resources like computing power and even conference rooms. Think of it as an eBay for internal resources that prices and allocates usage on demand.[12]

Huberman's system has a number of advantages over old-style reservation systems that allocate resources on a first-come first-serve basis. Reservation systems cannot accommodate new tasks as they arise, even if they are extremely urgent. When both demand and supply change constantly and unpredictably, as they do on a computer grid, markets perform much better. Imagine, for example, a business experiencing an unexpected surge in sales, or Web traffic that could purchase more resources as needed at a price set by the market.

After a successful internal trial run, HP's system has been handed over to CERN, the world's largest particle physics laboratory and a hotbed of grid-computing research, for further testing. Huberman hopes that the system might one day play a role in allocating computing power on a worldwide computing grid, where it will compete with similar solutions offered by IBM, Sun Microsystems, and others.

Corporate Communications

Corporate marketing and communications have traditionally been top-down activities. Most companies allow only a few key individuals to play a role in carefully shaping corporate images and brands. While some companies fire their employees for blogging on company time, smart companies are actively encouraging it.

Sun Microsystems CEO Jonathan Schwartz has been blogging for years. He may even be one of the first corporate executives anywhere to engage regularly in online conversations with employees, partners, shareholders, and customers.

When we asked Schwartz why he blogs he gave us an unexpected answer. He wasn't blogging for PR or to impress customers, or even to stroke his own ego. No. Blogging was just a more effective, more personable, and more transparent way of communicating with employees than sending an all-Sun e-mail.

"I wanted employees to understand why Sun executives were thinking the things we were, why we said the things that we did, and what more efficient vehicle than leveraging the network culture was really at the core of Sun to begin with," he said. Now Schwartz thinks that everyone at Sun should blog, and he has been actively encouraging more employees to take up the practice. "We're going to be driving unparalleled transparency into everything we do," said Schwartz, "precisely because it's the most efficient mechanism to accelerate change throughout Sun. Transparency enables everything to go faster, invites accountability, and drives dialogue between Sun and the communities we serve."

Not everyone is comfortable with the new dynamic, networked forms of communication.[13] "It is definitely alienating the old guard," said Schwartz. "They would like to believe that their groupwide e-mail is the exclusive vehicle for communicating direction." At the same time, Schwatz says blogging is definitely attracting a new guard at Sun. "It's yielding pace and transparency into our decisions and it's helping dissolve the boundaries between what is Sun and what is the market. And this in turn brings more and more people into Sun's ecosystem."[14]

Workplace Peering Is Here to Stay

These examples of how companies are taking unorthodox approaches to workplace functions such as teaming, time allocation, and decision making show that mass collaboration can succeed, even within the confines of a traditional firm. Indeed, self-organizing communities on the Web have proved time and again that they can be more effective in creating value than hierarchies—so why should it be different in the workplace? It is just a matter of shifting organizational paradigms. As self-organization becomes accepted as a viable method of production, more and more workplace processes will move from being hierarchically directed to self-organizing.

WAKING UP TO THE WIKI WORKPLACE

If an army marching in lockstep to tightly arranged military music is a metaphor for yesterday's workplace, the workplace of the future will be more like a jazz ensemble, where musicians improvise creatively around an agreed key, melody, and tempo. Employees are developing their own self-organized interconnections and forming cross-functional teams capable of interacting as a global, real-time workforce.

Such decentralization of the work flow and the actual workplace will be the defining trend in years ahead. Indeed, if Linux, Wikipedia, and other collaborative projects are any indication, it will often be easier and less expensive for workers to self-organize productively than to squeeze them into a corporate hierarchy.

While still in their infancy, these trends will change our experience of work, and especially the experiences of our children, deeply. Though these are long-term changes, self-organization in the workplace can give competitive advantage to firms that wield it effectively today. Loosening organizational hierarchies and giving more power to employees can lead to faster innovation, lower cost structures, greater agility, improved responsiveness to customers, and more authenticity and respect in the marketplace.

Is there is a danger that too much openness and self-organization in the workplace could lead to disorganization, confusion, and lack of focus and direction? Google CEO Eric Schmidt admits, "If you have worked in a traditional company, a place like Google doesn't feel right, it doesn't feel like you have the kind of control over the way in which decisions are made that you might have had in a more traditional environment." And yet, Schmidt is convinced that self-organization is better. "You talk about the strategy, you get people excited, you tell people what the company's priorities are, and somehow it works out," he says.

Clear goals, structure, discipline, and leadership in the organization will remain as important as ever, and perhaps more so as peer production emerges as a key organizing principle for the workplace. The difference today is that these qualities can emerge organically as employees seize the new tools to collaborate across departmental and organizational boundaries. Indeed, our research suggests that the results are often better when self-organization takes precedence. Many of Geek Squad's notable successes, for example, would not have come to fruition had Robert Stephens not loosened the reigns on the leadership to allow his employees to shine.

Where else might this new workplace modus operandi take us in the future? Predictions are always risky, but that's what books are for, so here are a few thoughts anyway.

New workplace environments. Is the centralized corporate headquarters a thing of the past? Many people, including John Seely Brown and Paul Duguid, have argued that they're not.[15] Proximity creates personal familiarity and fosters trust. Body language, intonation, and general demeanor play an important role in human interaction. Moreover, it's easy to underestimate the degree to which tacit learning and knowledge creation are enhanced by face-to-face contact.

These are all valid considerations, and will remain true for as long as we remain human. So while workplaces will not disappear (no, we will not all work from tele-cottages), there are fewer and fewer compelling reasons to organize monolithic physical workplaces to which the vast majority of employees report on a daily basis.

More than 40 percent of IBM's employees don't work in traditional offices—they work from home or on the road. At most companies the majority of employee communications already flow electronically through blogs, wikis, instant messaging, video conferencing, and various enterprise collaboration tools. As collaboration tools improve they will enable collaboration that looks and feels as though everyone is in the same room. The result is that workplaces will become smaller and teams will be more distributed, with participants drawn from all over the globe.

New economics of work. The days of lifelong employment and pensions are already long gone. But there is more change yet to come as firms seek to address the need for greater agility and lower costs. Employment relationships will necessarily become more fluid, definitely less long term, and undoubtedly more horizontal. Many employees will welcome this, as they search for flexibility, identity, ownership, authenticity, and continuous learning, both in the workplace and with their peers.

The creation of ad hoc, self-organized teams that come together to accomplish specialized tasks will become the norm. So look for consultancy to be the dominant contractual model for work in the near future, and expect more employees to demand a share of the profits derived from their intellectual contributions. Indeed, one of the big developments in the next decade entails a shift from voluntary and nonmonetary participation in peer-to-peer communities to a model where participants directly monetize their contributions. As this happens, we will see more and more freelancers, individual entrepreneurs, and SMEs taking on a larger and larger share of economic production.

New sources of identity and security. Our work may still largely define who we are, but employers no longer will. Our sense of stability and our sources of encouragement, learning, and growth in our careers will come from communities of practice and our engagement with like-minded peers who we meet and keep in touch with online, and not necessarily our long-term employment relationships. Rather, the people we meet at work join the personal networks we create as we move from organization to organization over the life span of our careers.

Expect new guildlike formations with codes of conduct that set the formal and informal norms and rules that govern how a growing number of people carry out their trade. Also, look for new peer-to-peer reputation-rating services to play a greater role in identifying high-quality, reliable collaborators.

New intermediaries in the talent market. Talent agencies, auctions, and markets will play a larger role in managing the interface between employers and employees. Human capital marketplaces like InnoCentive and collaboration brokers like CollabNet provide some noteworthy examples of the kinds of new intermediaries that will emerge as firms seek to interlace the contributions of people inside and outside corporate boundaries. Their value is in bringing a bit more structure and regularity to peer collaboration and making it easier for companies to tap into global talent pools on an as-needed basis. As companies learn how to harness these systems, traditionally vertically integrated corporate functions such as R&D and marketing will be radically transformed.

A DEMOGRAPHIC KICK FOR CHANGE

Don't expect overnight change. Previous shifts in organizational paradigms have been slow. The shift from cottage industries to the factory system unfolded over the better part of a century. The transition from industrial factories to today's high-tech office environments took at least a few decades.

One factor is institutional inertia. Organizations have their own internal logic, including rules, routines, norms, and power struggles. Practice shows that these intangible social elements of the workplace are much more difficult to change than the IT system. This is especially true of elders in the workplace, many of whom resist changes in their routines. The baby-boom generation grew up using typewriters, telephones, and cars to commute to work, and will have a difficult time changing its lifestyle. Technology may open doors, but it can't force people to walk through them.

There is no such problem with the Net Generation. "Look at the Net Gen," says Ross Mayfield. "You know, the children of the baby-boom generation who have five to seven instant messaging windows open at any time. When they're connected to the Net they're doing so socially. The computer is not a box, the computer is a doorway."

A truly self-organized and distributed way of working is not far off on the distant horizon. It's an imminent reality that few workplaces today are prepared for. "As the Net Gen enters the work space in really large numbers," says Mayfield, "there will be a very different, more informal method of working that this generation has already grown accustomed to. That will be a pretty big change in the workplace over the next couple of years."

Mass collaboration is already transforming the way goods and services are created throughout the economy, and it is now becoming a growing force in today's workplace. As a hypercompetitive economy combines with new technologies and the Net Generation, more companies are heeding the principles of openness, peering, sharing, and acting globally. Companies that embed these ideas in their workplaces will create competitive organizations that leverage internal and external capabilities more effectively than their traditional counterparts.

10. COLLABORATIVE MINDS
The Power of Thinking Differently

We began the book with the remarkable story of Rob McEwen, the visionary former CEO of Goldcorp who had the courage, and perhaps even the naivete, to challenge one of the mining industry's most deeply held assumptions: You don't share your proprietary data. McEwen saw things differently. Yes, geological data was important, but it was no use to anyone if Goldcorp's internal geologists were ill equipped to make sense of it.

When McEwen released the data on the Internet and challenged the world to do the prospecting, he transformed a lumbering exploration process into a modern distributed gold-discovery engine that harnessed some of the most talented minds in the field. With their varied experience and expertise, these contestants not only found gold, they introduced Goldcorp to state-of-the-art technologies and exploration methodologies, including new drilling techniques and data-collection procedures, and more advanced approaches to geological modeling. Harnessing these new technologies made Goldcorp an overnight center of excellence in Canadian mining and slashed the company's cost of production by over 60 percent in four years.

Though McEwen is too humble to admit it, his defiance of conventional wisdom turned out to be one monumental stroke of genius. The company soon stockpiled more gold than the Bank of Canada and then used its substantial war chest to acquire its largest competitor. By 2006, Goldcorp was North America's third-largest gold producer, and Red Lake mine remained the richest gold mine in the world.

The Goldcorp story raised a number of important questions for us that we have tried to address throughout this book. If a small, underperforming company in one of the world's oldest industries can achieve greatness by opening its doors to external input and innovation, what would happen if more organizations followed the same strategy? Couldn't just about any social or economic challenge be solved with a critical mass of self-organized contributors seeking an answer to the problem? In fact, wouldn't businesses be more productive if they could reach outside their walls to harness the insights and energies of a vast network of peers who converge around shared interests and goals? If so, how would the traditional corporation change? And what new business models could be built on this new collaborative approach to producing goods and services?

We've now traveled the vast terrain of the collaboration economy to discover that there are in fact several new models that companies can harness for greater competitiveness and growth.

- The peer producers apply open source principles to create products made of bits—from operating systems to encyclopedias.
- Ideagoras give companies access to a global marketplace of ideas, innovations, and uniquely qualified minds that they can use to extend their problem-solving capacity.
- Prosumer communities can be an incredible source of innovation if companies give customers the tools they need to participate in value creation.
- The New Alexandrians are ushering in a new model of collaborative science that will lower the cost and accelerate the pace of technological progress in their industries.
- Platforms for participation create a global stage where large communities of partners can create value and, in many cases, new businesses in a highly synergistic ecosystem.
- Global plant floors harness the power of human capital across borders and organizational boundaries to design and assemble physical things.
- Wiki workplaces increase innovation and improve morale by cutting across organizational hierarchies in all kinds of unorthodox ways.

Each model represents a new and unique way to compete, but they all share one thing: These new forms of peer production enable firms to harvest external knowledge, resources, and talent on a scale that was previously impossible. Whether your business is closer to Boeing or P&G, or more like YouTube or flickr, there are vast pools of external talent that you can tap with the right approach. Companies that adopt these models can drive important changes in their industries and rewrite the rules of competition.

So while the four principles of wikinomics—openness, peering, sharing, and acting globally—may actually sound pejorative to many managers, we have shown how they can be a powerful force to drive innovation and wealth creation. Not only that, the principles of wikinomics could help transform the way we conduct science, create culture, inform and educate ourselves, and govern our communities and nations. But in order to do so we must first confront some of the conventional business wisdom that has companies and other institutions mired in twentieth-century thinking.

In this chapter we describe some of the key business- and management-culture challenges that will need to be overcome to bring this new corporate worldview to fruition. How can leaders rewire their brains to think differently about the business world and resist the temptation to put up the barricades and fight the forces of self-organization in their industries? In essence, could all business leaders follow Rob McEwen's example and develop a collaborative mind?

Our experience suggests that this is much easier said than done. Indeed, the new collaborative culture engendered by Web 2.0 has provoked heated criticism from various observers, who worry that the democratization of value creation is having a corrosive effect on the institutions that have traditionally underpinned the economy and the production of human knowledge and culture. These critiques are worthy of debate, so our discussion of the power of thinking collaboratively begins with a rebuttal of what some critics have labeled the "dark side" of wikinomics. We argue that while some criticisms are arguably misguided, others point to important limitations and growing pains that accompany the shift to more collaborative and democratic models of peer production.

Second, we examine the crisis of leadership that is emerging as new business models threaten the old. This sheds light on the reasons why many companies find mass collaboration threatening and are choosing to dig in and fight instead. The media, entertainment, and telecommunications industries provide instructive examples, as we explore what it would mean to think with a collaborative mind.

THE "DARK SIDE" OF WIKINOMICS

Web 2.0 and the ideals associated with it are by no means universally celebrated. Whereas proponents (including ourselves) see the Internet's democratizing tendencies as a positive force that is broadening access to knowledge, power, and economic opportunity, its critics see it as a flattener of culture, an enemy of expertise, and a destroyer of wealth and property.

If the democratizing forces of the Internet are not counterbalanced, say some critics, human civilization is at risk of slipping into an untidy pool of mass mediocrity. Important distinctions between information and knowledge, fact and opinion, experts and amateurs, high and low culture will fade into history, as human knowledge and culture, and ultimately human civilization, spirals toward the lowest common denominator.

In probing the potential dark side of wikinomics, we examine two claims that have been championed by observers who remain skeptical about its benefits, if not outright hostile to its ideals. The first claim is that user-generated content on the Web is undermining the important role that gatekeepers play in maintaining high standards of quality, originality, and authenticity in media, entertainment, and culture. Crowds are more likely to produce ignorance than wisdom, say these critics. The second claim is that the Net Generation and their emerging "mash-up culture" is undermining intellectual property rights protection, while the proliferation of free content and services like Skype and Linux threaten to erode the capacity to generate wealth in a knowledge-based economy.

The Ignorance of Crowds

While it would be unfair and inaccurate to paint all Web 2.0 critics with one brush, there does seem to be at least one point that most critics agree on. Institutions need gatekeepers: people—generally highly credentialed people—who are entrusted by society to preserve the core traditions, values, and standards of practice that the institutions embody. Examples of such gatekeepers are plenty, from the publishers and editors of magazines, journals, and newspapers to the brand managers at leading ad agencies. In so far as the Internet undermines these traditional gatekeepers, critics argue that it is a largely negative force in today's society. Indeed, take away the gatekeepers, and all we're supposedly left with is a flood of amateur drivel.

As we alluded to in the preface, Andrew Keen's book *The Cult of the Amateur* exemplifies this school of thought.

> *So what, exactly, is the Web 2.0 movement? As an ideology, it is based upon a series of ethical assumptions about media, culture, and technology. It worships the creative amateur: the self-taught filmmaker, the dorm-room musician, the unpublished writer. It suggests that everyone—even the most poorly educated and inarticulate amongst us—can and should use digital media to express and realize themselves. Web 2.0 "empowers" our creativity, it "democratizes" media, it "levels the playing field" between experts and amateurs. The enemy of Web 2.0 is "elitist" traditional media.*

Keen's worry is that the rising tide of amateur creativity on the Web will drown out authentic talent, posing a grave threat to the vitality of culture and to the industries it engenders. The threat from blogs, wikis, and social networking is so dire, he insists, that it is pushing us back into the Dark Ages! No more Mozarts, Hitchcocks, or Hemingways. No more Universals, Warner Brothers, or Penguins. "If you democratize media, then you end up democratizing talent," says Keen. The unintended consequence of all this democratization is cultural "flattening." In the end we're left with nothing more than "the flat noise of opinion."[1]

Keen celebrates the old model, in which if you wanted to be a filmmaker, you had to go to the Hollywood studios. If you wanted to be a musician and be heard, you would go through the label system. If you wanted to be a published author, you needed to be signed by a publisher. The audience for these works, it seems, is simply incapable of identifying quality content in a world where traditional gatekeepers are losing power and authority.

Keen's argument is equivalent to saying that democracy is bad for the average citizen, because the average individual is a poor judge of his or her own interests. If the majority of the world accepted this principle, then most of us would still be landless serfs paying tithes to the feudal manor.

The Long Tail author Chris Andersen has rightly pointed out that the fantastic thing about the democratic systems of content creation on the Web is that they define talent and expertise much more efficiently than the old models did. "A lot of people are doing things that maybe wouldn't have passed the threshold or the test of admittance," says Andersen. "For instance, MySpace or YouTube are turning out to be tremendously popular, but they are not conventional." In the old model this content would have been ignored, because markets had limited shelf space, and you could only stock content that was guaranteed to be popular. "Now we have markets that have infinite shelf space," says Andersen, "so this content can get the audience it deserves." "We can offer everything," he adds, "and then measure what's actually popular."[2] If there is one thing that Keen is right about, it's that the new Web is inherently dangerous for the business models that depend on controlling the means of creation and distribution. And that is Keen's real fear. Elite artists and an elite media industry are symbiotic, he says—you can't have one without the other.

> *Traditional "elitist" media is being destroyed by digital technologies. Newspapers are in freefall. Network television, the modern equivalent of the dinosaur, is being shaken by TiVo's overnight annihilation of the thirty-second commercial. The iPod is undermining the multibillion-dollar music industry. Meanwhile, digital piracy, enabled by Silicon Valley hardware and justified by Silicon Valley intellectual property communists such as Larry Lessig, is draining revenue from established artists, movie studios, newspapers, record labels, and song writers.*

And yet, the conclusion that we are worse off as a result of Web 2.0 is fundamentally wrong. If anything, our cultural economies will grow to be more vital and more robust than ever. The opportunity to discover and nurture new Mozarts among a billion connected individuals is greater now than in a world where the fate of talented individuals depended on their ability to get noticed by an A&R representative from a major music label. The ability of consumers to collaboratively filter their way through the "noise of opinion" to get to the "good stuff " is already proven.

Perhaps the most important point to make in relation to Keen's thesis is that neither the old gatekeeper models nor the democratized Web 2.0 is perfect. Like the article in *Nature* comparing Wikipedia and Britannica, the real point is that traditional media and publishing can be just as bad as the worst of the Internet.[3]

Keen argues that the problem with Web 2.0 is that nothing is "vetted for accuracy." But as author and Stanford law professor Lawrence Lessig points out, even works that have supposedly met the high standards of truth imposed by publishers can be flawed, sometimes deeply. Making for delicious irony, he suggests that Keen's book is a perfect example of the limitations of traditional publishing.

"Here's a book that has passed through all the rigor of modern American publishing," says Lessig, "yet which is perhaps as reliable as your average blog post: No doubt interesting, sometimes well written, lots of times ridiculously over the top—but also riddled with errors." Lessig's point is that Keen's blind faith in the traditional system is misguided, and even dangerous. Gatekeepers make errors. And while the Internet doesn't primp itself with the pretense that its words are guaranteed to be true, traditional publishing is often held up as the pinnacle of truth and accuracy.

So what are we to conclude? "We need to understand the limits in accuracy, taste, judgment, and understanding shot through all of our systems of knowledge," says Lessig.[4] A challenge, we would add, that the democratized and fluid content creation mechanisms embodied in Web 2.0 are most readily able to tackle.

Take the recent efforts to improve the quality and accuracy of Wikipedia as an example. The online reference site enjoys immense popularity despite nagging doubts about the reliability of entries written by its all-volunteer team. A new program developed at the University of California, Santa Cruz, aims to help with the problem by color-coding an entry's individual phrases based on contributors' past performances.

The program analyzes Wikipedia's entire editing history—nearly two million pages, and some forty million edits for the English-language site alone—to estimate the trustworthiness of each page. It then shades the text in deepening hues of orange to signal dubious content. A one-thousand-page demonstration version is available on a Web page operated by the program's creator, Luca de Alfaro, associate professor of computer engineering at UCSC.

Other sites already employ user ratings as a measure of reliability, but they typically depend on users' feedback about each other. This method makes the ratings vulnerable to grudges and subjectivity. The new program takes a radically different approach, using the longevity of the content as an indicator of whether the information is useful and which contributors are the most reliable.

"The idea is very simple," de Alfaro said. "If your contribution lasts, you gain reputation. If your contribution is reverted [to the previous version], your reputation falls."[5] Over time, a user's history of edits is used to calculate his or her reputation score. The trustworthiness of newly inserted text is computed as a function of the reputation of its author. As subsequent contributors vet the text, their own reputations contribute to the text's trustworthiness score. So an entry created by an unknown author can quickly gain (or lose) trust after a few known users have reviewed the pages.

A benefit of calculating author reputation in this way is that de Alfaro can test how well his reliability scores work. He does so by comparing users' reliability scores with how long their subsequent edits last on the site. So far, the program flags as suspect more than 80 percent of edits that turn out to be poor. It's not overly accusatory, either: 60 to 70 percent of the edits it flags do end up being quickly corrected by the Wikipedia community.

While the program prominently displays text trustworthiness, de Alfaro favors keeping hidden the reputation ratings of individual users. Displaying reputations could lead to competitiveness that would detract from Wikipedia's collaborative culture, he said, and could demoralize knowledgeable contributors whose scores remain low simply because they post infrequently and on few topics.

Luca de Alfaro's efforts illustrate just one of many possible mechanisms that could help bring greater reliability and accuracy to the decentralized forms of content production that have flourished in a world where the means of creation are democratizing. Keen and other critics should stop pining for the old days and contribute to the conversation about how society can take advantage of the fact that talent and expertise are more distributed than at any previous time in history.

The End of Intellectual Property

We have argued throughout the book that digital technology and new models of collaboration are forcing a reexamination of intellectual property. We do not proclaim an end to intellectual property, but we do advocate the need to discover its new role as an enabler rather than as an inhibitor of cocreation and collaboration on the Web.

At the center of this transformation is the Net Generation. N-Geners like to remix and repurpose their media, their advertising, and their consumer items, and feel at liberty to share their remixes with the world. This inevitably creates intellectual property conflicts and clashes with a business culture that has stressed the importance of maintaining tight control over products and brands.

Scott Hervey, a veteran entertainment lawyer, says, "They want to be able to experience content without commercials, they are interested in finding new musicians that are not being pushed by major record labels, and they enjoy watching their contemporaries. They've got a different sense of entertainment. It's in much shorter bits and bites."[6]

Intellectual property hawks like the Business Software Alliance brand N-Geners as a generation of pirates and worry that they are increasingly indifferent to copyright law. Critics have been suggested that this generation's loose intellectual property norms could undermine the legal and economic foundations of capitalism.

But is it really piracy when DJ Copycat spends hours mashing-up the Sex Pistols' "Pretty Vacant" with tracks from The Charlatans and Visage into a new creation that he then shares with his followers online? Or is this exactly the kind of serendipitous creativity we should be encouraging as we look to invent new user-centric business models in the media industries?

Recently, a growing chorus of forward-thinking people and companies has been calling for lawmakers to roll back, or at least counterbalance, the ever-tightening regime of copyright law. The one thing this group holds in common is a lament for the efforts of entrenched Hollywood interests to further extend the life span of copyright protection. It's time, they say, to tip the balance of copyright law in favor of users and amateur creators.

Here's an excerpt from what veteran technology commentator Victor Keegan had to say in the *Guardian*:

> *We are now living in a digital age of instant and cheap availability, meshing and remixing, and of mass creativity, with increasing numbers of*

creators prepared to give their services free (as in much of the open source movement). We need fresh regulations for a new age before we cave into the demands of the producers as they try to get draconian rules put in place before the shutters come down on the old world.

The creative economy is vitally important, but the way to nurture it is to follow the winds of the information revolution and not the desire of existing corporations to preserve a business model that has been turned upside down by the revolution taking place in virtually every creative industry.[7]

Keegan's right; we need a level playing field on which new forms of cocreation and innovation can flourish. If Hollywood titans succeed in locking up their vast archives of media content for another twenty years, it would certainly be a setback for open content initiatives, but it will most certainly not be a death knell. The regulatory measures deployed by incumbents amount to temporary stopgaps, not sources of enduring competitive advantage. In any case, a more balanced approached to the way we recognize, encourage, and reward knowledge production will be driven by viable new business models that can fuel value creation in intellectual-property intensive industries, not by changes in the law.

The law is a blunt instrument and, when it comes to the regulatory arena, concentrated corporate interests usually win out over the diffuse interests of users. Incumbent companies already use existing patent and copyright protections as a shield against innovation and change in their markets. And, for as long as new business models remain elusive, media behemoths will exploit the heavy hand of regulation to enforce consumer behaviors that are in fact contrary to consumer interests.

That doesn't mean that we shouldn't lobby for user-friendly copyright policies, but we shouldn't rely on changes in the law to bring about a more decentralized, user-centric media environment either. Ultimately the most effective antidote to the media industry's political maneuvering is to demonstrate that the explosion of amateur creativity on the Web can be a source of immense social, cultural, and economic value. If media companies wish to participate in this value creation they will need to quickly come to terms with the fact that they can hold more in an open hand than they can in a closed fist.

For an example of how traditional companies and content creators can effectively leverage Web 2.0, consider Lucasfilm's efforts to enable users to create their own *Star Wars* movies. Rather than trying to prevent fans from creating and sharing videos related to the popular movie series, users are being encouraged to become codirectors (although Lucasfilm takes extensive copyright ownership of the subsequent creation). The newly redesigned *Star Wars* Web site empowers fans to "mash up" their homemade videos with hundreds of scenes from *Star Wars* movies; watch hundreds of fan-made *Star Wars* videos; and interact with *Star Wars* enthusiasts from around the world like never before.

Some may see this as revolutionary, and in relation to what most other companies are doing it is. But really, it's just common sense, particularly when you consider the marketing value, community engagement, innovation, and fan loyalty that the initiative is generating. The real question shouldn't be why Lucasfilm is doing this, but rather why most other companies and content creators aren't following suit.

WHEN WORLDS COLLIDE

"What if I don't want to open source my intellectual property?" you may be asking. Certainly, many in Hollywood feel that way. Second Life developer Linden Lab gives all property rights to users, and even encourages secondary markets in characters, objects, and "land." But larger studios such as Sony disagree. Whatever is created inside the game belongs to Sony, and they have fought customers who attempt to sell virtual game goods on eBay.

Reticence in the face of monumental change is not unusual, and in some cases it's entirely rational. Take the publishers of media, music, software, and other digitized goods who have not simply stuck their collective head in the sand; they have a very legitimate business problem: Publishers can't reasonably adopt open approaches that would cannibalize existing revenues without a viable means to shore up their ailing income streams. Jim Griffin, managing director of OneHouse LLC, calls it "Tarzan economics." "We cling to the vine that holds us off the jungle floor," he says, "and we can't let go of the one we've got until we've got the next vine firmly in our hand."[8]

The problem is that media incumbents are moving too slowly. They're getting mired in the thick underbrush of thorny contractual agreements and outdated and costly infrastructures. What's worse is that the economic foundation of the industry is based on a business model suited for the era of analog publishing, not for a world of user-driven creation and distribution. These institutions are powerful and deeply ingrained in the industry's social and economic contract. It's hard for senior executives to imagine a world where their companies could lose control of the very resources they have monopolized for so long.

That's why the publishing industry has always liked the "information superhighway" metaphor for the net. They see the net as one big content-delivery mechanism—a global conveyer belt for prepackaged, pay-per-use content, not a platform for peer-to-peer collaboration. In order for this vision to work, however, publishers need to exert control through various digital rights management systems that prevent users from repurposing or redistributing content.

Competing with "Free"

Smart companies are pioneering new business models that begin with an entirely different premise: users, not distributors, are in control. That premise follows from the realization that the opportunity to control the quantity and destiny of bits is vanishing along with the last remnants of the analog recording industry. So the next strategic frontier is to build business models that compete with the free content that consumers can download from the Internet. For Nettwerk Records, a growing Canadian record label, the key to success in an industry reshaped by consumer-controlled distribution is to monetize the one thing you can't commoditize—the emotive value of the music experience.

Nettwerk Records started in Terry McBride's Vancouver apartment in 1984. Today it's Canada's largest independent, boasting artists such as Avril Lavigne, Sarah McLachlan, Ron Sexsmith, the Barenaked Ladies, Sum 41, and Dido. Although McBride supports copyrights, he does not tow the usual industry line on intellectual property rights. He says the solution is not to stem the flow of technology, but to understand which technologies consumers are using, and then find creative ways to make music part of the experience.

If kids are listening to music while playing video games, McBride wants Nettwerk involved. If people are downloading songs while text messaging,

McBride wants Nettwerk music to be available. He's not particularly fussy about the medium—the content is his message. McBride's recipe for success is short and simple: "We're going to support every platform and format that comes, and we're going to allow the fan to consume our music however they want to consume it. We're not going to tell them how to consume it. . . . In other words," he adds, "music should be like water: let it flow in every way possible."[9]

Music flowing like water? It sounds reminiscent of the cyber-libertarian refrain: "Technology wants to be free." But McBride has established a serious track record in the business, making him a force to contend with. He claims to have had the same knee-jerk reaction the rest of the business had when file-sharing technologies like Napster emerged. "But once we started understanding it, getting educated on it and talking to the fans, we realized this is a whole different ball of wax, and there is a way to make this work." McBride seems to concede that copies of his artist's music will inevitably circulate on the Internet for free. But he's bullish about the ability of Nettwerk to find creative ways to monetizing the fans emotional connection to the music his artists produce.

Fans of the Barenaked Ladies, for example, can buy the recording of the concert they just saw. "Copies of the live show are available within five minutes of the concert ending," says McBride. During a tour of the UK in 2007, USB live sales were about 70 percent of the Barenaked Ladies' media sales. McBride estimates that 5 to 10 percent of the people who showed up at the show bought a USB stick before they went home. "That's a huge conversion rate," he says. "We're usually lucky if we get a one percent or two percent conversion rate on CD sales."

Nettwek has also experimented with increasingly open models of music production. When it came time to produce a Sarah McLachlan Christmas remix single, for example, the traditional approach would have been to spend thirty to forty thousand dollars hiring a producer to do the remix. McBride figured that that was too much to pay for a single that would sit in the marketplace for six weeks, so he organized a remix contest on a major DJ site. Hundreds of DJs submitted entries, and thousands of fans voted for their favorites. McBride purchased the top three remixes and started to sell them. "It cost us nothing besides the file purchase price, which was only a couple of thousand dollars," says McBride, "and we made more than what we spent on just the digital sales within a two-week window."

Opening up the process to external DJs also led to copious free promotion. "I know for a fact that every DJ that spent a lot of time and effort to make their mix would have played it in their own sets at the clubs that they were playing at," says McBride, who figures that the same prosumption approach could be applied to everything, from T-shirts to album cover designs. "There's many different ways to harness the emotional connection with your audience and involve them in the creative process."

McBride's open approach appears to be working. Online sales account for 40 percent of Nettwerk's sales, which are in the tens of millions, compared with the 10 percent to 15 percent his competitors average. Now some of the major labels are beginning to follow suit. For example, EMI, the third-largest music company by sales, decided to offer up their catalog DRM-free, both on Amazon and iTunes. Will EMI's artists start outselling artists from other labels? And if so, will the other big music companies come around? Even better, will some artists demand that their labels do the same, and/or choose EMI if given a chance?

It's too early to say just yet, but EMI's leaders have put down their marker. "The recorded music industry . . . has for too long been dependent on how many CDs can be sold," says Guy Hands, EMI's chairman. "The industry, rather than embracing digitalization and the opportunities it brings for promotion of product and distribution through multiple channels, has stuck its head in the sand."[10]

In the end, EMI's move has the ring of the inevitable. Most technologists have long agreed that DRM is a lost cause—hackers reverse-engineer it just as fast as it gets produced. As Fred von Lohmann of the Electronic Frontier Foundation explains, "In a connected world, all it takes is one person anywhere on the planet to break a DRM system and extract the contents—from that point forward the content circulates freely without further restriction. It's a break once, break everywhere problem. A DRM system that's not perfect is useless."[11]

Besides that, DRM is simply bad for customers. It limits their freedom to enjoy content on the devices and formats of their choice. And in a world where customers rule, that means DRM is bad for business. Most major players are reluctant to accept this, however. And the result is that new business models for open content aren't likely to be pioneered by traditional media conglomerates, but by independent artists like Radiohead, tech companies such as Google, and smaller, more nimble labels like Nettwerk.

This new generation of companies is not burdened by the legacies that inhibit the publishing incumbents, so they can be much more agile in responding to customer demands. Radiohead, a band that has long spurned industry norms, figures that if fans are going to download their music anyway, they may as well make it easy to download their songs legally. Their most recent album, *In Rainbows*—released in October 2007—is only available through Radiohead's Web site. But Radiohead did something else that few artists or labels have dared: They let customers decide what to pay for it. Contrary to conventional wisdom, estimates suggest that millions of fans paid an average of between 6 and 8 dollars for the album. Since Radiohead is independent, they don't share that revenue with a label. It sure beats the average of 72 cents that the typical major-label artist earns from a CD sale of $19.99.

Overcoming the Innovator's Dilemma

What a growing number of artists seem to understand is that you don't need to control the quantity and destiny of bits if you can provide a superior value proposition for customers. Free content is a reality that is here to stay. So artists will have to provide customers a product that is better than "free"—a product that is more convenient, more compelling, and that provides more value than what customers can get from illegitimate sources. For successful companies and artists, the ability to provide exceptional value is their hallmark.

Radiohead provided its fans with a second CD of additional photos, artwork, and lyrics, all packaged in a "discbox." Artists like Nine Inch Nails and the Beastie Boys have provided compelling venues in which fans can build communities around sharing and remixing their content. Apple's iTunes competes with "free" every day, and it does a very good job. As von Lohmann points out, "For the customers that find iTunes useful, they find it more convenient; there are millions of people every day who are using iTunes instead of downloading it from a file-sharing network."[12] The same could be said of Netflix, the popular online DVD rental and video streaming service. All of the videos provided by Netflix are readily available on Bit-Torrent networks for those who put in the effort to find them. But loyal customers stick with Netflix because of the flat-fee business model and its innovative customer-driven recommendation and rating service. The fact that innovation and success in figuring new business models for the entertainment industry has been driven by nontraditional players underscores a key point. Traditional firms threatened by self-organizing prosumer communities face the innovator's dilemma.[13] The innovator is able to attack markets with low-quality, low-cost options that a market leader would never consider. Additionally, in the case of prosumer communities, the innovator taps a low-cost or no-cost volunteer production resource.

With some ingenuity, the publishing and entertainment industry could get itself out of this mess. EMI's Guy Hands says, "Radiohead's actions are a wake-up call which we should all welcome and respond to with creativity and energy." Roger Faxon, chief executive at EMI, agrees. "The commercial roles of music companies will be more as facilitators for bringing music and the rights that support them into the market place," he says, "as opposed to being originators of the content itself."[14]

In addition to new business models like those developed at Nettwerk, the industry as a whole should consider mechanisms that transfer a percentage of revenue from content aggregators such as YouTube and My-Space back to all of the hardworking people who are directly and indirectly employed in the production of great content. Alternatively, Jim Griffin has suggested a broad, but very small, levy on Internet connections that could be redistributed to artists using actuarial monetization (much like royalties distributed to musicians by publishing societies such as ASCAP, which periodically sample radio play sheets in order to determine them). This would create a real incentive for artists to contribute to a global media commons while still providing a fair way to compensate them for their work. Our favorite is a subscription model, in which consumers get access to an unlimited selection of music tracks through streaming Internet audio for some kind of recurring payment, much the way satellite radio operates today.

Solutions like these need fleshing out. We are optimistic, however, that the majority of media companies will eventually see the light. Those that don't will go the way of the sheet music publishing business of the early twentieth century that failed to adjust to the rise of the recording industry.

WAR ON THE OPEN INTERNET

When it comes to the Internet, Hollywood and the telecommunications companies are brethren—they both face the innovator's dilemma, and they both see the Internet as a wild beast that needs to be tamed.

Telecommunications is in a state of disarray. In a world of free Internet telephony, a major source of revenue is set to disappear altogether. Upstarts like Skype are ascendant, and though landlines will not disappear overnight, the writing is on the wall. Telephony will be free. It's just a matter of time.

Telecommunications companies are understandably unwilling to hasten the transition by offering free phone service in an attempt to wipe out Skype. But failing to do so could mean losing the ability to compete with Skype in the future. Skype already operates at a mere fraction of the cost of a traditional telecommunications carrier.

There is no doubt a strong hint of irony when the very baby that telecommunications firms like AT&T helped to birth is now a source of great pain and upheaval. Like the media industries, telecommunications firms face a genuine business problem. They need to recoup their investments in maintaining and upgrading the telecommunications infrastructure. But with free services cutting into their revenue, they have deemed the service fees they charge consumers and businesses for Internet connections too paltry by way of compensation. So they want to create a tiered Internet with different levels of service, akin to first class, business, and coach.

William Smith, chief technology officer of BellSouth, has already proposed charging fees in exchange for giving one Web vendor's traffic priority over that of a competitor. If it pays the freight, BellSouth users will find Yahoo's search engine works faster and better than Google's. So, in effect, BellSouth becomes a gatekeeper for the types of services that will thrive on the Internet—an Internet where bandwidth and content-delivery rights are auctioned off to the highest bidders.

This poses a grave threat to the Internet—a threat that could extinguish the fire of innovation that has spurred countless new business, including most of the examples discussed in this book. This is not just a war against the open Internet; it's a war against economic development, a war against competitiveness, and a war against innovation. In short, it's a war against the future. Internet pioneer Vint Cerf reminds us that the remarkable social impact and economic success of the Internet is in many ways directly attributable to the architectural characteristics that were part of its design. The three golden rules—nobody owns it, everybody uses it, and anybody can add services to it—are what distinguish the Internet from any previous communications medium. "By placing intelligence at the edges rather than control in the middle of the network," says Cerf, "the Internet has created a platform for innovation. This has led to an explosion of offerings that might never have evolved had central control of the network been required by design."[15] Indeed, services such as Skype, Google, flickr, Linux, MySpace, and Wikipedia might still be just twinkles in their creators' eyes.

There is a great irony in what the telecommunications lobby is doing. These free services are in fact the reason why customers are upgrading to broadband. "Providing tiered service is not good strategy," says Google CEO Eric Schimdt. "It could materially and negatively affect the adoption rate."[16]

In the end, people and technology move forward. And just as sure as the ocean's waves wash away beach sand, companies that choose to stand still will be swept aside by the momentum. "You cannot prevent the inevitable," says Schimdt. "You may be able to slow it temporarily, but the slowing just makes the ultimate endgame come faster."[17]

THE CRISIS OF LEADERSHIP

There has probably never been a more exciting time to be in business, nor a more dangerous one. The wikinomics genie has escaped from the bottle, wreaking havoc on some and bestowing long-term success on those who embrace it.

This is a paradigm shift. Paradigms are mental models that constrain our thinking and are often based on assumptions so strong we don't notice them. New paradigms cause disruption and uncertainty, even calamity, and are nearly always received with coolness, hostility, or worse. Vested interests fight against the change, and leaders of the old are often the last to embrace the new. Consequently, a paradigm shift typically causes a crisis of leadership.[18]

Look around and you can see how most firms are slow to respond to the mass collaboration revolution. Like the publishers and telcos, they underestimate the threat, and by the time they adapt it's too late. So the key is to fine-tune your radar and move quickly to seize the opportunities for influence as a new business paradigm takes hold.

If Sun and Microsoft had launched an aggressive attack on Linux in 1996, for example, they might not have won, but they could have changed the future landscape. They didn't do that, and today they're busy adopting open source tactics in their business as fast as they possibly can. If everything else fails, companies can do as some telcos are doing and fight back with laws and regulations. This does not attack the superior value proposition of innovators like Skype, but it may barricade them from entering your market.

The best route is to move early to participate in building a brand-new business. By joining the community of new innovators you might even position yourself at the front of the parade. This is what Red Hat and IBM have done with Linux. Joining early gives you the ability to influence major issues, such as the strategic direction, standards, and rules. If it's late, you can align yourself with the community to prevent damage from future battles.

The lesson of history is that profound changes favor the newcomer and, in rare cases, the incumbent firms that learn to think differently. Value will migrate to new players the way that the telegraph business lost out to telephones or the way that PCs displaced the mainframe. But that is the nature of the capitalist beast. Holding back technology to preserve broken business models is like allowing blacksmiths to veto the internal combustion engine in order to protect their horseshoes.

Sure, ask any manager if they would prefer to compete in a "well-mannered" economy, in which every new innovation was subject to their approval, and they would overwhelmingly respond in the affirmative. But a well-mannered economy is not today's reality.

Stability is dead. The idea that you can invent a business that will never be disrupted by technology is over. As blogger and science-fiction author Cory Doctorow put it, "Blacksmiths weeping into their beer about their inability to sell horseshoes in the era of railroads doesn't make horseshoes more popular. Blacksmiths learning how to become auto mechanics, on the other hand, puts food on their table."[19]

The choice facing firms is not whether to engage and collaborate with peer-production communities but determining when and how. Now that people have access to tools for creation and distribution they will use them—for their own ends and on their own terms. New disruptions—Wikipedia, flickr, open source software, and blogging—are emerging all the time. The opportunity for customers or competitors to get the jump on new innovations in your area of business increases daily. And with the speed at which these communities move, the time to act is now.

11. ENTERPRISE 2.0
Harnessing the Power of Wikinomics

In February 2007 Swiss drug maker Novartis did something rather unusual—and almost unheard of in the high-stakes, highly competitive world of big pharma. After investing millions of dollars trying to unlock the genetic basis of type 2 diabetes, the company released all of its raw data on the Internet—for free. This means anyone (or any company) with the inclination can use the data without restrictions.

Type 2 diabetes and related cardiovascular risk factors—including obesity, high blood pressure, and high cholesterol—are among the most common and most costly public health challenges in the industrialized world. Pinpointing their precise genetic origins could unlock a treasure trove of new medicines and result in a major windfall for Novartis's shareholders.

So why the giveaway? Mark Fishman, president of the Novartis Institutes for BioMedical Research, says, "These discoveries are but a first step. To translate this study's provocative identification of diabetes-related genes into the invention of new medicines will require a global effort."[1]

In other words, the research conducted by Novartis and its partners at MIT, Harvard, and Lund University in Sweden merely sets the stage for the more complex and costly drug identification and development process. According to researchers, there are far more leads than any one lab could possibly follow up on alone. So by placing its data in the public domain, Novartis hopes to leverage the talents and insights of a global research community, to enable it to dramatically scale and speed up its early-stage R&D activities. Novartis also points out that open collaborations help boost the morale of the company's scientists, who relish the idea of establishing tighter relationships with their colleagues in academia.

It's worth noting that Novartis didn't reveal everything. For example, it didn't give away its own observations on the data after nearly three years of deep research, which means it retains a substantial lead time on other companies that attempt to exploit the data.

Meanwhile, the close ties and goodwill that Novartis has fostered with the research communities studying diabetes will give it an advantage over companies that lack these relationships, as it moves to the next stage of research. As Susan Gasser of the Friedrich Miescher Institute in Switzerland pointed out to *The Scientist*, "If an independent investigator has results that will help cure diabetes, he or she might eventually come back to Novartis to collaborate—since academic labs cannot do drug development."[2]

HARVESTING THE BUSINESS VALUE OF WIKINOMICS

The Novartis initiative nicely encapsulates the principles of wikinomics.

Novartis is open. When it defines the boundaries of the enterprise, Novartis thinks not just about the ninety-eight thousand people it employs full-time, but about a broad array of individuals and partner organizations in industry, government, the nonprofit sector, and academia that can enrich its value proposition.

Novartis peer produces value by collaborating closely with these other institutions in society to achieve its business goals. And rather than keeping everything secret, Novartis shares some intellectual property in order to foster relationships and stimulate progress in the research communities that conduct the basic science that contributes to the pharmaceutical industry's growth.

Finally, Novartis thinks and acts globally in order to bring new drugs to market. While still relatively young (Novartis was born of a merger between Sandoz and Ciba-Geigy in 1996), it operates in 140 countries, and its research institutions stretch from their headquarters in Cambridge, Massachusetts, to Basel, Switzerland, to Singapore, where it operates one of the world's largest centers for research on tropical diseases. Novartis also considers its responsibilities as a global citizen; in addition to research on malaria, tuberculosis, and dengue fever, Novartis has an access-to-medicines program through which it donates 2 percent of its total net sales to disadvantaged patients around the world.

In embracing these four principles—openness, peering, sharing, and acting globally—Novartis is transforming itself into an increasingly collaborative enterprise model we call "Enterprise 2.0." Enterprise 2.0 is a new kind of business entity, one that opens its doors to the world; co-innovates with everyone, especially customers; shares resources that were previously closely guarded; harnesses the power of mass collaboration; and behaves not as a multinational but as something new: a truly global firm.

We have cited plenty of evidence throughout this book to support our thesis that firms that embrace the Enterprise 2.0 model can harness external resources and talent and achieve unparalleled growth and success as a result. The hard part, as we argued in the previous chapter, is rewiring your brain and turning off those old business reflexes so that you can capitalize on what the new world of wikinomics can offer.

To aid your efforts to internalize this new business paradigm we'll now review some of the key lessons and examples of how smart firms are putting the principles of wikinomics into practice.

Being Open

A growing number of smart companies are learning that openness is a force for growth and competitiveness. As long as you're smart about how and when, you can blow open the windows and unlock the doors to build vast business ecosystems on top of what we call platforms for participation.

Amazon, eBay, Google, and flickr open up their applications and business infrastructures to increase the speed, scope, and success of innovation. Their platforms for participation create a global stage where hundreds of thousands of customers and partners add value and establish synergistic businesses. Amazon, an open platform prodigy, harnesses the power of two hundred thousand active developers and gains nearly 30 percent of its revenue from third-party sellers that leverage its e-commerce engine.

Not just Web-based companies. Perhaps you're thinking that openness is easy for Web-based companies but not for companies in other parts of the economy. Think again. P&G was notoriously secretive and about as closed as a company can get. It didn't look outside its walls for anything, and it certainly didn't let anyone look in. In early 2000 the company had a near-death experience. Its business lines were atrophying, and revenues and profits were down as a result. The stock price had tumbled, and Wall Street was screaming. Something had to be done.

Putting more money into internal R&D wasn't attractive—its innovation success rates were flat lining. So A. G. Lafley (CEO) and Larry Huston (innovation head honcho) led the company on an ambitious campaign to restore P&G's greatness by sourcing up to 50 percent of its new innovations outside the company. In addition to broadening and deepening its own proprietary networks, P&G searched for innovations in marketplaces such as InnoCentive, NineSigma, and yet2.com. These combined efforts led to hundreds of new products on the market, some of which turned out to be hits. In the process Lafley and his managers, like Huston, transformed a lumbering consumer products company into a limber innovation machine.

Today P&G is a leader among thousands of companies that participate in a global ideagora, where uniquely qualified minds exchange millions of ideas and inventions in something akin to an eBay for innovation. Companies that move now can leverage a global pool of talent, ideas, and innovations that vastly exceeds what they could ever hope to marshal internally. Remember, P&G figures that for every top-notch scientist inside its labs, there are another two hundred outside who are just as good.

How can you get with the program? The starting point for any manager is personal use of the new collaborative technologies, preferably in conjunction with a Net-Generation youngster. Ask your college student son or daughter to show you Facebook. Join MySpace. Edit a page in Wikipedia. Create a video clip for YouTube. Get a taste for how these new open communities work.

The next step is to start a planning process with a comprehensive map of your innovation ecosystem that positions your value creation and assesses the interdependencies that will determine the flow of benefits and your ability to capture a share of them. This is not a traditional competitive landscape or value-chain analysis but an analysis of the participants creating knowledge pertinent to your existing and future business. While this includes business partners and competitors, it extends to academia, public research institutes, think tanks, creative communities or communities of practice, and contract research organizations. The map needs to be global and cover all relevant disciplines that intersect with your strategy.

This will help resolve some important questions about openness. Where are your competitors innovating, and how should you focus internal resources to ensure that you can continue to differentiate yourself in the marketplace? Are employees plugged into the right knowledge-creating networks? What products, processes, or assets (like Amazon's e-commerce platform) could you open up to reduce R&D costs, increase growth, and grow the number of participants in your ecosystem? Or which ideagoras should you join to source key innovations or even license-out innovations of your own (i.e., a specialized marketplace like Eureka Medical or a broader marketplace like yet2.com)?

Timing and tactics. Our research shows that the timing of opening up products and platforms is important. Most companies find that a staged evolution toward openness provides the right balance between controlled growth and innovation on one hand and self-organization on the other. Facebook is a case in point, where each evolution toward increased openness proceeded on a firm foundation of past success.

The increasingly popular social network began in 2004 as a network for connecting college classmates and keeping in touch with friends from school. Soon 85 percent of all North American college students were hooked. In September 2006, Facebook opened its membership to anyone who wanted to join and quickly grew to over thirty million members. Then, in May 2007, it invited its users and third parties to add new social computing applications (called widgets) to the Facebook platform.

Today, Facebook users can assemble their own personalized toolbox of widgets from a collection of thousands of applications, much the way you can customize Firefox with thousands of third-party extensions. No longer content to be just a social network, Facebook is vying to become an "operating system for the Web," with which users can do just about anything they currently do online, with the added benefit of being able to incorporate their social network seamlessly into the experience. Everything from planning a vacation and shopping to reading the news and watching sports can now be done in the company of one's Facebook network.

Following in Amazon's footsteps, Facebook has also promised that developers and service providers will have significant freedom to monetize their widgets through ads and transaction revenue. Given its demographics, this makes Facebook a pretty potent business engine for small and large companies alike.

No tax, no tariffs competition. Now, not every company wants to be an open platform, and not every company should be. Apple didn't want to open up the iPod or its iTunes service. It's fighting the French government, which has launched a legal challenge to force Apple to make music purchased from iTunes' competitors playable on the iPod. The French fear that a Microsoft-like monopoly is in the making, while Apple argues that it shouldn't be compelled to open the playing field to competitors and that doing so could compromise the exceptional user experience that Apple has built through tight integration between the iPod and the iTunes service. Opening the platform to customer innovation, on the other hand, could create more value for all customers and further enhance the versatility and popularity of the iPod. Perhaps then openness wouldn't seem so threatening to Apple, and both the French and Apple would be happy.

It goes to show, in any case, that openness isn't easy. Think back to what Shai Agassi, formerly SAP's product and technology group president, said about their decision to open up thirty thousand APIs to its enterprise-software platform: "It's almost like you're taking down your borders and opening for no tariffs, no tax competition. You need to know that your core assets and your skill sets allow you to continue to innovate fast enough as a corporation."[3]

If you're going to be naked, you better be buff. Of course, openness cuts both ways. In the brave new world of transparency, companies have less and less control over information and consequently less control over the perceptions of their firm and its products. Individuals can inform others, shape public discourses, and ultimately alter the decisions and actions of individuals and groups—often with little more than a blog post. Dell recently learned this lesson the hard way, when the computer maker was blindedsided by an ex–Dell employee, who submitted a piece to The Consumerist (a popular consumer advocacy blog) entitled "22 Confessions of a Former Sales Manager." The article revealed a few tricks about how to get the best deals and support from Dell. It received almost three hundred thousand hits and much public attention in the span of three days.

Like the proverbial deer caught in the headlights, Dell made a classic blunder. The legal department deemed that the article contained "confidential and proprietary" information. It hastily asked The Consumerist to take the post down. Instead of removing the post The Consumerist posted Dell's demand, which was subsequently viewed by 130,000 people and dugg 3,500 times on Digg.com. In a face-saving decision, Dell posted a retraction of their demand along with their own confessions list, admitting to their mistake of trying to control information. Even the largest companies, it seems, have little choice but to yield to the wishes of these powerfully networked online communities. Perhaps more important, the episode demonstrates that openness is the best policy. As Lionel Menchaca, Dell's digital media manager, later put it, "Ignoring negative issues is not a viable strategy in the blogosphere. If you aren't prepared to discuss negative issues head-on and actually fix what's causing the negative conversations, be ready to fail publicly."[4]

Embracing openness. Despite this recent misstep, Dell has gone further of late than many companies in embracing openness as a business philosophy. Dell's Direct2Dell blog hosts candid discussions from employees on Dell's customer service, its products, the technology behind them, and the technology industry in general. Direct2Dell has brought a new level of openness to the company's product quality and its customer service by giving customers a direct line to actual Dell employees (not offshore service reps), who spend time responding to readers' comments on the blog.

Dell's finest effort is IdeaStorm (www.dellideastorm.com), an innovative Web site where customers provide advice on how to improve Dell's products and services. Launched in February 2007, IdeaStorm looks and feels a lot like Digg.com—users post suggestions and the community votes, so that the most popular ideas rise to the top. As of March 2010, the site had hosted over thirteen thousand that have been voted over seven hundred thousand times. Admittedly, not every idea is great. And some communities speak louder than others. But Dell has taken customer input seriously, and now provides a full summary of "ideas in action" on its site. This sends a signal to the community that their feedback is being heard. Among other things, Dell is offering preinstalled Linux on its PCs, in response to demand from a large contingent of open source enthusiasts. CEO Michael Dell told us that he sees customer-driven innovation like this as the linchpin of his strategy for Dell 2.0. "We need to think differently about the market and engage our customers in almost everything we do," he says. "It's a key to us regaining momentum as a technology industry leader."

Openness does up the ante—it drives real value to the fore and forces every company to compete on a level playfield. But the alternative—closing your doors to external input, scrutiny, and participation—is much less compelling. Like P&G in 2000 you may find yourself isolated from the networks that are creating value in your field and unable to keep up with rising marketplace demands for growth and innovation.

Peering

IBM joined the Linux peer producers and gave away hundreds of millions of dollars of software and resources to support them. Did IBM lose its head? No, it stumbled onto a new mode of production called "peering" that will harness human skill, ingenuity, and intelligence more efficiently and effectively than anything we have witnessed previously.

For all intents and purposes the Linux community is an extension of IBM's human capital. But IBM cannot control what Linux developers do. You could even argue that one of its most important assets lies outside its corporate boundaries. But IBM saves bundles on development costs and generates billions in revenue from Linux-related services and hardware every year. Linden Lab harnesses peer production by enabling its customers to cocreate Second Life in a very meaningful and ongoing way. It produces less than 1 percent of its game content and instead gives powerful scripting tools to its members. Virtually every character, object, and experience in Second Life is created by thousands of enterprising residents, who exploit the intellectual property rights in their creations to participate in a thriving virtual economy with a $100 million turnover.

Self-organized projects like Linux and Second Life marshal the efforts of thousands of dispersed individuals, sometimes in miraculous ways. Loose, voluntary communities of producers can self-organize to do just about anything: design goods or services; create knowledge; assemble physical things; or simply produce dynamic, shared experiences. But don't overlook the fact that these communities operate according to well-defined norms and have internal structures and processes to guide the group's activities.

Anarchy, hierarchy, or meritocracy? In its early days Linux developers were characterized as idealists, outsiders, hackers, and anarchists. The popular media tended to portray open source projects as an amateurish free-for-all, with no central direction or individual in charge. Well-heeled opponents of open source software took advantage of these perceptions to make a case for why open source is incapable of yielding the quality, innovation, integration, and completeness of proprietary software. All successful open source communities today deploy structured and hierarchically directed processes for managing the tedious, tiresome work of joining together all of the fragmented pieces and contributions. This balance between self-organization and hierarchical direction allows these communities to harness an incredibly diverse talent pool while still achieving the tight integration required for something as sophisticated as an operating system.

Though Linux relies on contributions from thousands of programmers, a core group led by Linus Torvalds make strict judgments about which contributions of code make it into the kernel of the operating system. "I certainly support anybody's right to modify and publish their own version of Linux," says Torvalds, "but at the same time, almost all my efforts are spent on the actual joining back of the results, and that's what most of what you'd call 'core developers' end up doing: guidance and quality control at various levels."[5]

The importance of community coordination and integration cannot be overstated. If value is not integrated properly, users may become disenchanted and stop contributing. Open source programmers report dropping out of projects that miss their release dates too frequently. Some of the value that contributors produce must be locked up in the successful accomplishment of goals.

Benevolent dictators. Like any leader of a meritocracy, Torvalds is not immune from challenges, and has been forced to respond constructively to many criticisms from within the community. This sort of benevolent dictatorship characterizes many open source communities. Issues are debated publicly on e-mail lists and Web sites, which helps build consensus for final decisions on, for example, which code and features to include in official releases of programs. Debates remain in the public archive. The meritocratic structure of the community supports the growth of open source. Community code sites and sympathetic press sites enable rapid diffusion of updated code and spread news of open source developments to the greater information technology community. This role is crucial. Charismatic projects get good press and attract programmers. Conferences are especially important. Programmers often work alone, and individuals from many countries may contribute to a project. Bringing people face-to-face builds the trust the community needs to iron out disagreements.

The broader message for companies wishing to harness peer production is that effective communities need leadership and structure, typically at many levels of the community and at various critical junctures in the production process. This potentially leaves the door open to individuals and companies who can brings technical skills, management skills, and other forms of specialized knowledge or know-how that is not already abundant in the community.

Any company seeking to open source a product or participate in peer production communities, for example, could help devise control points and collaborative processes for weeding out poor contributions and assembling end products. Remember, companies can add a lot of value in the commercialization process, including development, testing, marketing, and distribution. It may make sense to do this internally, since much of the proprietary differentiation from open source products will be achieved in the way the companies integrate, deliver, and support their offerings with a variety of complementary services.

Developing synergies with a peer community. For another example of how companies can build synergies with large communities of peers, consider the incredible success of Mozilla, the producer of the Firefox Web browser, an open source pioneer and a leader in applying the principles of wikinomics to product development and marketing. In the years since Firefox was released it has single-handedly broken the Internet Explorer

(IE) monopoly, benefited all Web surfers by reviving browser innovation, been downloaded over a billion times, and captured roughly 2 percent of the browser market—a remarkable set of achievements for a small non-profit organization.

In most respects Mozilla is more of a movement than a corporation. Its users are religious about its qualities, and this grassroots enthusiasm translates into one of the most effective peer development and marketing models on the planet. After all, Firefox competes with a threadbare budget, versus the $500 million Microsoft recently spent on launching its new Vista browser. Whereas Microsoft employs an army of hundreds of IE programmers that toil away in Redmond, the Mozilla Foundation manages the Firefox code with only a dozen full-time developers. You'd think this might tilt the play field in Microsoft's favor, but Mozilla relies on a volunteer community with hundreds of thousands of participants who are committed to making Firefox better.

About four hundred trusted contributors make regular additions to the code, usually after a two-stage review. Thousands of additional contributors submit software patches to the core team (often to flaunt their skills in hopes of being recognized). An even larger group downloads the full source code each week to scrutinize its performance as the software modules evolve. Finally, more than five hundred thousand people are pegged to test run beta versions before each new Firefox iteration goes live (about one fifth of them take the time to report problems and offer to help fix bugs). Altogether, this makes Firefox powerful, user-friendly, secure, and, above all, dynamic, because it is always improving.

Mozilla's equally impressive peer-marketing efforts are something to behold. Since 2004, its most passionate adopters congregate at SpreadFirefox.com—a "seeding" ground for evangelists, who promote and support community development and localization efforts around the world. Among other things, evangelists can participate in a number of exposure-building programs, including Get Firefox, a referral program boasting some of the world's top 250 referrers, and Firefox flicks, a user-generated, content-driven site where users post thirty-second video testimonials expressing the virtues of the open source browser.

The engagement of the Firefox evangelist community goes deeper still. A group of Oregon students recently made a forty-five-thousand-square-foot crop circle of Firefox's logo. Another advocate tattooed the logo onto his head. And incredibly, over fifty thousand donors paid to underwrite a two-page ad in the *New York Times*, spurring further Firefox adoption and buzz. Similar fund-raising programs have been developed in other Firefox geographies around the world.

In this vast hive of activity, the Mozilla Foundation is merely a catalyst. Asa Dotzler, Mozilla's head of community development, says, "It starts with a genuine Firefox mission to build, above all else, the best experience for the Internet user possible. We then provide the infrastructure, leadership, and support to empower users to get involved in this cause."[6]

Like Linux, the Firefox community is a "meritocratic hierarchy." "The more you engage," says Dotzler "the bigger your reputation, the more you're given access, and the more you're listened to." This self-governing community system feeds the curiosity, passion, status-seeking ego, and sociability of its "ambassadors," who as nonpaid members of the Mozilla community can approve new developments, speak to the press, and host parties on behalf of Mozilla. In fact, a large percentage of Mozilla Foundation's two hundred employees have been plucked from experts, enthusiasts, and hackers who first resided within their evangelist communities.

Providing the tools for mass collaboration. Remember that mass collaboration increases when the tools for creation and consumption are widely distributed and the goods in question are nonrival (i.e., my consumption of the good does not deplete the supply available for you to consume). By increasing the degree to which a good is nonrival, or by increasing the tools available to end users, one can spur self-organized production. There are many examples of firms using both strategies. When Facebook opens its platform to users and third-party developers, it is distributing tools for production. When Novartis releases its genomic data on type 2 diabetes, it sends a message that this data set is a nonrival good.

Fostering community spirit can help sustain peer production efforts over time. Mozilla goes to great lengths to foster community spirit, through local events and parties and by immortalizing contributors and evangelists on the Firefox Friends Wall. Mozilla has also established a set of guiding principles for decision making that puts the needs of its user and developer communities at the heart of its decision calculus. As the organization's leaders consider key opportunities and challenges, they filter each decision through a series of questions.

1. Will our developments truly benefit our users (as opposed to merely benefiting our company)?

2. Will it enrich the user's experience and not mess up what's already working for them?

3. Does what we're doing have consensus among our community of users?

4. Do we always behave from a position of consistent values with our audience?

5. Have we maximized the amount of authenticity and transparency to deliver participation, accountability, and trust?

6. Are we letting our passion show, even if this means that we're occasionally wrong because of it?

7. Do we start by directing the following questions to our audience "what do you think and why?" and "what value could we provide you?"

Developing the capacity to work with peer production communities effectively is not something that happens overnight. The Mozilla Foundation was steeped in open source from the beginning. But most large corporations are accustomed to a much more formal and hierarchical way of working. IBM has honed its skills in working with the Linux community over more than a decade of experience and relationship building. Firms that are lagging in peer production will find this capability difficult to imitate.

Sharing

Smart firms today understand that sharing is more than playground etiquette. It's about lowering costs, building community, accelerating discovery, and lifting all boats in the sea. Indeed, the increased willingness of countless individuals and organizations to share is giving rise to a rich legacy of public goods on the Web. Computing pioneer Dan Bricklin calls this the "cornucopia of the commons": The growth of increasingly valuable online resources as a natural by-product of their use by self-regarding individuals (i.e., no altruistic sharing motives need be present, especially since sharing is the default in most online communities, including flickr, Napster, and YouTube).[7]

In a growing number of cases, sharing can be orchestrated to great effect. We saw the California Department of Education, for example, harness the power of sharing by open sourcing its textbooks, putting high-quality educational materials within every student's reach and saving taxpayers $400 million per year.

Organizations like the Bill and Melinda Gates Foundation and the Tropical Disease Initiative are leveraging open source drug discovery to launch an unprecedented attack on neglected diseases such as cholera and African sleeping sickness. Similar open source initiatives are appearing across the sciences, when researchers share their data, tools, results, and computing resources in massive cross-institutional collaborations.

The business value of sharing. But what about the business of sharing, you ask? Let's revisit the Skype discussion from the previous chapter. We know it's causing grief for the telecommunications business and provoking its leaders into a backlash that could end up ruining the Internet. But consider Skype's positives.

eBay saw Skype as a valuable tool to achieve improved communication between its members. It would make the eBay experience better, and ostensibly lead to more trading. But having paid $2.6 billion to acquire the company, eBay obviously saw more business potential than that. Like Google, Microsoft, Yahoo, and others, eBay bet that Internet telephony is the future, and further that there will be plenty of opportunities to bundle other services around it, even if voice is a free commodity.

While customers may not pay for the calls, they will pay to make calls to the landline network, for voice and video messaging, and for other things that haven't been invented yet. The trick is to determine what consumers will pay for. And that's where Skype has an advantage—it doesn't bear the burden of legacy business models. Indeed, it can experiment for very little cost and at very little risk on the Web, and in ways that incumbents can't.

Managing your portfolio of IP. So how can you avoid ending up like the telcos, and harness the power of sharing in your business? Start by balancing your IP portfolio. Selectively open up some assets as a means to spur innovation in your ecosystem. Then catch a ride on the next disruptive wave. Sometimes you'll find that valuable follow-on innovations will come from customers, the way Lego's Mindstorms products became more valuable after the company opened up the source code. Other times breakthroughs will come from collaborators in a community of practice, the way Intel harvests contributions from a broad network of university scientists to get a jump-start on innovations in the field.

It helps to think about how you treat knowledge resources within your organization, and across the broader business ecosystem of which your firm is a part. What information assets do you have that would gain value if they were treated as public goods? Are you treating digital goods as rivalrous when they are nonrivalrous? And how does this stance affect the opportunities for innovation? As Amazon CEO Jeff Bezos has said, "I think it is something every company can do, if they look inside and think what are some unique assets that others might enjoy."[8]

Remember: Just as good personal-investment strategies diversify financial assets across a range of low- and high-risk opportunities, good innovation strategies diversify intellectual assets across a range of open and closed offerings. Sharing makes sense if the conditions are right, so here are some rules of thumb and examples from earlier chapters.

- Your proprietary offering is falling, and open sourcing it will inject the creativity and manpower required to help it succeed in the marketplace (Sun's OpenSPARC initiative).
- Opening up IP in one area of your business will boost demand for complementary products and services offerings (IBM's WebSphere product).
- The advantages of pooling competencies and reducing R&D costs exceed the benefits of having exclusive rights in the knowledge produced (SNP Consortium).
- You're seeking a uniquely qualified mind, so you enlarge the pool of talent addressing a particular problem (Goldcorp).
- An open platform will encourage innovation, efficiency, and interoperability with ecosystem partners (Amazon).

- Sharing is necessary to establish credentials and to develop relationships with other contributors in the community (SpikeSource).
- Preempting the property rights of competitors shifts the locus of competition or enhances your freedom of action (Merck's Gene Index).
- Openness removes unnecessary friction in collaborative projects and paves the way for the participants to focus on the task at hand, rather than on the property rights (Intel's industry-university partnership).

Even with these guidelines, a lot of the art and science of sharing revolves around hiving off those elements of value creation that you own and monetize versus the value that is collectively owned by the community. This gets tougher when more people participate in the ecosystem, and if anything, open-licensing entities such as the Creative Commons can ensure that media content, software code, and other digital artifacts remain open for use, modification, and redistribution.

Economic incentives for mass collaboration. And yet, there is a wide spectrum of options that fall between "completely open" or "completely closed." IBM director of strategy Dan McGrath says, "The interesting model is the one that falls in the middle ground—a model that says we're collaboratively building something that will be privately owned by our consortia, or maybe your shares of ownership are apportioned in proportion to how much you brought to the table." One could envision a "digital-age co-op" with peer-rating systems that dynamically apportion shares to contributors based on the community's assessment of the value added by individual contributors. Annual profits from sales and services could then be distributed across the community of contributors. Whatever the precise arrangement, it's clear that the future of mass collaboration lies in hybrid models, in which participants share and appropriate at the same time.

Indeed, when business leaders first encounter wikinomics, they often wonder what motivates people to contribute to mass collaboration. The answer is ultimately as complex as human nature itself, but the motivations essentially boil down to a mix of intrinsic and extrinsic rewards. As discussed previously, most examples of mass collaboration are fueled by the personal utility that individuals derive from their participation in the creative process, including intrinsic rewards like the satisfaction of having solved a tough problem or the desire to meet unique needs that commercial firms have not addressed.

The importance of intrinsic rewards is backed up by research. In a survey of nearly seven hundred open source developers conducted by the Boston Consulting Group, programmers reported that intrinsic rewards such as creativity and autonomy outweighed extrinsic rewards such as income, career advancement, and skills development. This was true even if the contributor was paid by her employer to work on an open source project.

While sufficient to entice to collective production of many noncommercial goods, intrinsic motivations by themselves are probably not sufficient to sustain collaborations where the principal objective is to make a profit. Quite simply, it's hard to maintain any sense that "we're all in this together" if one party disproportionately benefits from the voluntary efforts of others.

Companies seeking to tap wikinomics should consider how the formal and informal incentive systems they provide will affect the success and longevity of their initiative. So far two monetary incentive frameworks have emerged: contests and microeconomies.

Contests are one way to reward people who contribute to products and services. Design contests, programming contests, and marketing contests have become increasingly popular, as big name brands from Converse to Sony to Frito-Lay seek out new ways to infuse customer input into their product strategies and marketing campaigns. Take Frito-Lay's "Crash the Super Bowl" contest. Interested parties were invited to submit thirty-second spots for the Doritos brand, and the broader community on the Web was invited to vote for their favorite submission. The winning ad aired during the Super Bowl. Some entrepreneurs are betting that contests will figure prominently in some industries and are setting up specialized intermediaries.

One company staking a claim in the contest economy is TopCoder, which organizes some of the world's largest computer programming competitions. Such competitions have been around for decades, but growing worldwide demand for exceptional talent has upped the ante, and competitions are growing in size and frequency. Google's recent "Code Jam" attracted 20,000 participants from 143 nations, up from 14,000 in 2005.

For the contestants in these tight-knit community contests, cash prizes are often secondary to the potential for recognition that could jump-start a career. TopCoder claims it can be a "write your own ticket" kind of event. For competition sponsors, the contests provide a low-cost, low-risk way to identify and test out new talent.

Contests aren't perfect, though. Critics question the ethics of contests, in which thousands sign over the rights to valuable knowledge, while only a handful of winners take all. If a contest sponsor launches a multimillion-dollar service based on contest output, then somewhere in the region of twenty-five thousand dollars in prize money looks like a pittance.

While these challenges are real, TopCoder chairman Jack Hughes says that although 3 to 4 percent of the one hundred thousand members grab a large share of the rewards, 20 percent of their contest participants have sold some code through the network. Still, the shelf life of such contests appears to be shrinking, and the winner-takes-all model doesn't foster community building or collaboration among competitors, so a careful balance must be struck.

The rise of microeconomies. The most promising incentives structures revolve around microeconomies, in which user contributors have an opportunity to profit from their contributions to the community. From eBay to Amazon to CaféPress to Second Life to YouTube to Facebook, microeconomies are driving the most exciting and advanced incentive-distribution structures. To capture value from this evolution, companies need to create ways to distribute rewards among the community of co-innovators. CaféPress and eBay have already done it, YouTube and Facebook are trying to figure it out, and many more are on their heels.

Acting Globally

Like others in the aerospace and defense industry, Boeing has found the costs, risks, and expertise required to engage in large-scale development projects like designing and building new aircraft are simply too large for it to go it alone. So Boeing reached beyond its walls to cocreate the 787 with a network of partners that stretches over six countries. It gave up control over a large proportion of the thousands of components that make up its airplane and sacrificed some critical engineering knowledge. In the process Boeing is transitioning to a smaller, more capable workforce that is honing new skills to manage a globally distributed team made up of different companies, cultures, and disciplines. Has Boeing made the right long-term gamble?

We believe they have. The best-in-class knowledge and capabilities to design and assemble a high-tech jumbo jet today are spread over hundreds of firms around the planet. So Boeing is replacing its old hierarchical producer-supplier relationships with a global network of partners that share the risks and rewards of designing and manufacturing new aircraft. Close collaboration throughout the life cycle of the project enables Boeing to harvest the best of its partners' ideas. In exchange, Boeing's partners participate in the project's upside. In the end, everyone comes out with a better product for less cost and in less time than had Boeing opted to keep its partners at arm's length.

Going global. Like many of its contemporaries, Boeing is moving beyond the multinational model to something new—a truly global firm that breaks down national silos, deploys resources and capabilities globally, and harnesses the power of human capital across borders and organizational boundaries. But what's particularly new about the wikinomics era is that small firms can act global too.

We saw, for example, how self-organizing Chinese motorcycle firms are threatening the Asian export markets that Honda, Yamaha, and Suzuki once dominated comfortably. Honda saw its share of the substantial Vietnamese market fall from 90 percent to 30 percent in five years, and now Chinese motorcycle manufacturers like Lifan are making inroads into North America and Europe. At the same time, the proliferation of low-cost fabrication capabilities means that small and medium-size enterprises can make and sell products to a global market without having to manufacture anything themselves. New services such as Ponoko, based in New Zealand, can arrange to have your products manufactured and delivered direct to the customer, virtually anywhere in the world. All one needs to do is to upload your design to the Web site and select the materials—Ponoko does everything else. Entrepreneurs who are just getting started can even post their products to Ponoko's marketplace. Chief strategy officer Derek

Elley says, "It's a bit like low-cost global manufacturing and peer-to-peer commerce straight from your living room."[9] Creators can now turn their ideas into tangible offerings with less risk, lower costs, instant scalability, increased control, and less complexity. In turn, consumers get lower prices for individualized products, and perhaps most important, the proposed manufacturing model promises to reduce the environmental impact tied to production—mostly through the elimination of intermediaries and a reduced need for transportation. All considered, these forces will eventually lead to increased micromanufacturing, and even personalized manufacturing, as digital fabrication technologies make us genuine producers of the everyday objects that have long been the province of large-scale industrial manufacturers.

Outsourcing is more than arbitrage. Another key theme that emerged from our research is that becoming a truly global enterprise means abandoning the view that outsourcing is just a way to off-load costs. Outsourcing is increasingly a way to gain speed, innovation, and knowledge. Whether you're large or small, great global companies fashion their products in a way that exploits the latent knowledge capital that resides in its web of suppliers on one end and its retail partners and customers on the other. Indeed, suppliers, retailers, and customers become more like integrated partners, as everybody in the business web shares information through interenterprise systems that enable the ecosystem to act as a single entity. Remember, the global plant floor begins with the creation of the product content, and connects all the way down to the ultimate consumer, with the sharing of data, knowledge, and expertise all the way through.

Sounds radical, but as our IDC colleague Bob Parker has argued, if you have ever played (or observed someone playing) a massively multiplayer role-playing game (MMPRPG) like Second Life, World of Warcraft, or Ultima Online, you have seen the future of manufacturing business webs. In these games players assume certain roles (i.e., magician, blacksmith, etc.) and self-organize to conduct campaigns that mitigate the collective risk and share the spoils (i.e., gold, experience). Similarly, massively multiorganization business webs (MMOBW) will bring together knowledge workers from many industry companies all over the globe to design, make, and sell products. These MMOBW will take intelligent factory automation to the next level, with technology that enables the acquisition and delivery of real-time data, as well as the ability to remotely control the factory equipment to initiate corrective action. Don't worry: Your next generation of workers, who have grown up playing video games, will know how to do this, and how the information interfaces should look.

Global citizens. On the other end of the spectrum, we have explored how in numerous cases citizens can also act globally. A surprising number of affluent North American families, for example, are taking advantage of sites like Elance to outsource personal tasks to freelancers in India, China, and Bangladesh. Jobs such as landscape architecture, kitchen remodeling, and math tutoring, and small tasks such as getting a dress designed, creating address labels for wedding invitations, or finding a good deal on a hotel room, are being done for a fraction of what they'd cost at home. The *Wall Street Journal* estimates that such "personal offshoring" still represents a tiny fraction of the more than $20 billion overseas outsourcing industry, but claims that it's likely to evolve into a larger niche as offshore workers identify the opportunities.

Meanwhile, we got a vivid reminder of how global citizens have the potential to reshape politics and the media one evening in July 2006, when the evening news was full of images and sounds coming from the war between Israel and Lebanon. We surfed around the Internet to get a better read on what people on the ground were saying. We soon discovered that ordinary citizens were posting some of the most captivating (and at times disturbing) footage of the conflict on YouTube—not just frightened faces next to bombed-out buildings, but the real, unedited stories of ordinary Israeli and Lebanese people who got caught up in the conflict's misery.

While the Iraq war has citizens and soldiers using blogs and digital photos to document their experiences, the violence in Lebanon and Israel ignited just as video-sharing sites such as YouTube were capturing the mainstream imagination. YouTube visitors could browse through hundreds of graphic, and often passionate, citizen accounts of the chaos, destruction, and suffering spawned by the daily clashes between Hezbollah and the Israeli army. In many cases citizen journalists captured the experiences of their families, friends, and countrymen with a veracity and intimacy that major networks could not replicate.

So get ready for the hyperempowered citizen. The new generation of digital citizens has the means of creation at their fingertips, so that anything that involves information and culture is grist for the mill of self-organized production. Their expectations of business and government are higher, and like Geek Squad their philosophy of work revolves around fun, diversity, flexibility, and peering. They are not consumers but prosumers. And they have grown up in a globalizing world. In fact, a growing number tend to think of themselves as global citizens first and foremost.

These digitally shaped expectations represent a far more radical shift than previous generation gaps. They entail a fundamental rethinking of how every institution in society operates. Are today's senior managers listening? The Net Generation is knocking.

WIKINOMICS BY DESIGN

So how should leaders go about applying the principles of wikinomics in their businesses? Knowledge management theorist David Snowden says you should throw away some of your detailed plans. He thinks effective leaders manage chaos the way a kindergarten teacher manages her students. "Experienced teachers allow a degree of freedom at the start of a session, then intervene to stabilize desirable patterns and destabilize undesirable ones," he says. "And when they are very clever, they seed the space so that the patterns they want are more likely to emerge."[10]

In a similar way, your planning must allow for a high degree of learning on your part, and the flexibility to respond to new opportunities that arise out of the interplay among participants in your business web. Mass collaboration is a design-and-production innovation, and the firm must learn how to operate in this new environment. Most of the great examples we have seen, from IBM and Linux to Second Life and Amazon, have adopted their strategies as they learned what worked and what did not work for the overall community. Across the seven models of mass collaboration, however, there are several additional design principles that are common to most if not all of them.

Take Cues from Your Lead Users

Flickr's meteoric success is linked to its knack for cultivating a deep-rooted social community that is intensely involved in the evolution and growth of the site. Interestingly, flickr started in 2003 as an online multiplayer game in which the photo-trading feature was almost an afterthought. And yet that offering took off, effectively changing the direction of the company. Soon hundreds of thousands of members were uploading, annotating, and commenting on each other's shots, establishing tight-knit communities of friends and colleagues around a shared interest in photographic content.

Build Critical Mass

The secret to successful peering is building a critical mass of participants that attracts more and more people to the ecosystem. In fact, few projects survive without ongoing funding and support and a core group of leaders that provides guidance and manages interactions. These actors provide the social capital and technical infrastructure that other participants build on. Just as many open source software projects are to a large degree the creatures of Linus Torvalds and Richard Stallman, for example, the ongoing development of Wikipedia relies heavily on the vision, values, and commitment of Jimmy Wales and a core of Wikipedians. To extract long-term benefits from peer production communities you need to help create and replenish this critical mass.

Supply an Infrastructure for Collaboration

An important part of creating critical mass involves cooperating to supply the open standards, shared IP, legal foundations, and collaborative infrastructures that will support the innovation process. The Apache Web server, for example, would be nowhere if not for the leadership and resources IBM dedicated to help establish the Apache Software Foundation (the independent legal entity that shepherds the community's activities and protects its legal and economic interests). And yet, the investment needed to sustain resources like these could be small compared to the benefits that companies receive. Better still, the costs could be spread across a group of companies that collectively stand to benefit, as in the SNP Consortium.

Take Your Time to Get the Structures and Governance Right

The SNP Consortium took over a year to flesh out its model of collaborative research and development. This included convincing the leaders of individual companies to set aside their proprietary efforts to collaborate in an open fashion. Once there was agreement on the premise, the participants haggled over a legal framework that set out the right division of labor and funding; they recruited university research centers to work on the project; and they built an infrastructure for collaboration with the support of two major IT vendors, including a shared set of computing resources that were required to carry out the genomic research. In the end the collaborative research effort more than repaid the time and effort required to get the processes right by dramatically lowering their costs and speeding the industry toward a new era of personalized medicine.

Make Sure All Participants Can Harvest Some Value

People who participate in peer production communities have all of kinds of reasons for jumping in. Some do it for fun or philosophical reasons, and others do it for profit. Some participate to fulfill unmet needs, and still others seek the reputational benefits that can translate into better career prospects. These diverse motivations mean the community stewards must pay special attention to the balance of public and private activities. Using a fine-tuned approach to allocating property rights over the outputs of collaborative innovation can help manage the tensions that arise in cocreation environments. Providing the right for everyone (including free riders) to enjoy noncommercial benefits keep the barriers to participation low. Reserving the right to appropriate private returns to those who make substantial contributions, on the other hand, will reward those who put in the most effort. Communities can use this more granular approach to ensure that all contributors in the ecosystem can obtain some benefits.

Abide by Community Norms and Create Conditions for Trust

Individuals involved in mass collaboration may have highly individual motivations and goals, but they are expected to follow rules and protocols. There are written and unwritten rules that govern issues such as communications, appropriation, and the form and manner of contribution. Wikipedia uses community norms to guide editorial decisions. The site says it explicitly aims to be unbiased, and to present all points of view sympathetically, and judges contributions by that standard. Community norms have also been influential in shaping Wikipedia's funding policy. Jimmy Wales has backed away from adopting an advertising model (despite the potential for a windfall) for fear of alienating the community, which has so far resisted the idea. Firms that want to participate in peer production need to understand and abide by the norms that govern the community.

Let the Process Evolve

Danny Hillis, who founded Thinking Machines and invented parallel computing, says there are two ways to build complex things: engineering and evolution. IBM chose evolution when it decided to join rather than fight peer production. The firm did study open source and set appropriate goals to guide its activities, but it learned how to succeed through trial and error. IBM strategist Joel Cawley cautions, "Keep in mind that there was no strategy to do what we did with open source. It was happenstance all the way through the journey. We started doing what made sense, and kept on doing that each step of the way."[11]

Don't Lose Sight of Your Business Objectives

Wikinomics is not a mission, a goal to be achieved for its own sake. Business leaders should never abandon control of their destiny. Nor should businesses give up all of their core intellectual property. Wikinomics *does* mean having well-developed and well-understood internal goals to guide external engagement strategies. If firms can't realize these goals, they shouldn't invest their resources. And remember, openness, peering, and sharing are not panaceas to solve all business problems but complements to proprietary approaches.

Collaboration Starts Internally

When confronted with such an all-encompassing revolution in business most companies wonder where they should begin. Our answer is to start at home, by fostering collaboration in the workplace. After all, it's hard to collaborate with external partners without first getting effective collaboration happening internally. As discussed in Chapter 9, the advent of Web 2.0 tools—weblogs (blogs), wikis, shared bookmarks, and tagging—along with some repurposed older tools like forums allow companies to engage in a more vibrant and multiway communication inside their organizations. Early adoption programs could focus on using social networking technologies to uncover hidden sources of expertise inside your organization and generate strong bonds within and between teams. Wikis provide a perfect venue for collaborative brainstorming, project development, and project documentation and management. Blogs can serve as ways to foster an internal discussion culture, in which employees at all levels are encouraged to have a voice, while corporate leaders can blog to transform the way they communicate the vision, values, and priorities of the firm to customers, partners, and employees.

Finding the Internal Leadership for Change

All new business paradigms face challenges: Who gains and who loses? How difficult is the learning curve? Where will the time come for the organization to make use of this? Why should we invest in an initiative our part of the business did not demand? For every enterprise, the mix of these (and similar) factors is simply a part of working life. Ignoring the cultural memes and norms of the organization is a common way to have the principles and practices of wikinomics discredited at initiation. Champions of this kind of change need to be sensitive to these concerns and structure their engagement with the enterprise in a way that allows the benefits of collaboration to manifest themselves in pilot projects that can be scaled up and help build further momentum for cultural change.

Hone Your Collaborative Mind

Engaging in collaborative communities means ceding some control, sharing responsibility, embracing transparency, managing conflict, and accepting that successful projects will take on lives of their own. This can be awkward for companies accustomed to command-and-control systems. It means learning new skill sets that emphasize building trust, honoring commitments, changing dynamically, and sharing decision making with peers.

Years ago the major hotels cooperated to create a common reservation network. According to Gordon Kerr, vice president of MIS for Hyatt Hotels, at the time it took about a year "to understand that it was in our customers, and therefore our, best interest to do this." Each hotel chain had to avoid temptations to tilt the system slightly in their favor. "We had to each shut off the 'compete at all costs' part of our brains to work for the common good." Again, this is not about being nice—it's about rewiring your brain for the new paradigm of competitiveness.

WIKINOMICS AND BEYOND

A new force in business is emerging. We've called it mass collaboration. Linux, MySpace, and Wikipedia may have captured the popular imagination, but mass collaboration goes way further. It's a new way for people to socialize, entertain, and transact in self-organizing peer communities of their choosing. Companies can design and assemble products with their customers, and in some cases customers can do the majority of the value creation. Scientists can reinvent science by open sourcing their data and methods to offer every budding and experienced scientist in the world an opportunity to participate in the discovery process. Even governments can get involved, by using the new digital collaboration tools to transform public service delivery and engage their citizens in policy making.

For the business manager, the number one lesson is that the monolithic, self-contained, inwardly focused corporation is dead. Regardless of the industry you compete in, or whether your firm is large or small, internal capabilities and a handful of b-web partnerships are not sufficient to meet the market's expectations for growth and innovation.

Winning companies today have open and porous boundaries and compete by reaching outside their walls to harness external knowledge, resources, and capabilities. They're like a hub for innovation and a magnet for uniquely qualified minds. They focus their internal staff on value integration and orchestration, and treat the world as their R&D department. All of this adds up to a new kind of collaborative enterprise—an Enterprise 2.0 that is constantly shaping and reshaping clusters of knowledge and capability to compete on a global basis.

Managers should treat wikinomics as their playbook and harness its core principles to achieve success. The new age of mass collaboration will no doubt seem complex and uncertain, and it's true that collaboration and openness are more art than science. Leaders must prepare their collaborative minds. And companies will need unique capabilities to work in collaborative environments. Capabilities to develop new kinds of relationships, sense important developments, add value, and turn nascent networked knowledge into compelling value are becoming the bread and butter of wealth creation and success. Is your mind wired for wikinomics?

ACKNOWLEDGMENTS

Many important business thinkers and practitioners contributed to this book. Individuals from member companies who supported our *IT & Competitive Advantage* research project and who contributed important insights include:

Pat Adamiak, HP; Miley Ainsworth, FedEx; Pierre-Paul Allard, Cisco; Lynn Anderson, HP; Terry Assink, Kimberly-Clark; Jim August, IBM; Gery Barry, Blue Cross and Blue Shield of Louisiana; Dan Belanger, Service Canada; Bill Belew, Citi (retired); Bridget Bisnette, Cisco; Tracey Black, RBC; Alison Blair, Ministry of Health, Province of Ontario; Larry Blakeman, MetLife; Shawn Blevins, SAP; Brett Bonner, Kroger; David Bonskey, DaimlerChrysler; Chris Brennan, Service Canada; Mark Bruneau, BCE; Amy Buck, P&G; Jimmy Burk, FedEx; Rob Carter, FedEx; Joel Cawley, IBM; James Cortada, IBM; Dan Delmar, IBM; Nick Donofrio, IBM; Bill Dowell, Herman Miller; Martin Duggan, Service Canada; Suzanne Eger, Disney; Stacey Fish, SAP; Maryantonett Flumian, Service Canada; Douglas Frosst, Cisco; Bruce Harreld, IBM; Wendy Hartzell, GM; Alan Hibben, RBC; Kevin Hill, P&G; Neal Hill, Cognos; Martin Homlish, SAP; Paul Huebner, Ameritas Acacia Companies; John Hummel, Sutter Health; Kevin Humphries, FedEx; Jon Iwata, IBM; Tom Kegley, Roche Diagnostics; Fred Killeen, GM; Stanley Kubis, Kimberly-Clark; Jonathan Landon, Kimberly-Clark; Laurie Laning, P&G; Dan Latimore, IBM; Edmundo Llopis, Citi; Paul Logue, HP; Don Ludlow, RBC; Steve Mann, SAP; Kris Manos, Herman Miller; Bill McDermott, SAP; Bill McEwan, Sobeys; Ken Naigus, Disney; Deb Nelson, HP; Filippo Passerini, P&G; Kal Patel, Best Buy; Ruchi Prasad, Nortel; Mike Prosceno, SAP; Ann Purr, LOMA; Val Rahmani, IBM; Catherine Rein, MetLife; Ron

Ricci, Cisco; James Richardson, Cisco; Dan Schutzer, Citi (retired); Robert Scott, P&G; Tony Scott, Disney; Effie Seiberg, IBM; Steve Sheinheit, MetLife; Steve Slattery, Nortel; John Smith, Canada Post; Steve Stewart, FedEx; Barbara Stymiest, RBC; Ralph Szygenda, GM; Rich Taggart, Disney; Steve Tennyson, TD; Karenann Terrell, Baxter; Mark Thomas, FedEx; Fred Tomczyk, TD; Paul Tsaparis, HP; Sue Unger, DaimlerChrysler; Sam Uthurusamy, GM; François Vimard, Sobeys; Kase Waddell, Disney; Joe Weinman, AT&T; Irving Wladawsky-Berger, IBM; Andy Wnek, Canadian Tire; Bill Wohl, SAP; Debbie Wolter, formerly AT&T; Bill Zeitler, IBM.

A number of people spent nontrivial time working with us on this book, contributing their insights. For this we are very thankful. They include:

Shai Agassi, SAP; Brad Anderson, Best Buy; Jeff Barr, Amazon; Alf Bingham, InnoCentive; Mike Bair, Boeing; Tim Bray, Sun Microsystems; Tantek Çelik, Technorati; Jennifer Corriero, TakingITGlobal; Steve Delaney, Celestica; Cory Doctorow, Boing Boing; David Flaschen, Castanea Partners; Michael Furdyk, TakingITGlobal; Jim Griffin, One House LLC; Aden Holt, Buzz Oven; Larry Huston, P&G; Pradeep Jotwani, HP; A. G. Lafley, P&G; Larry Lessig, Stanford University; Barry Libert, Shared Insights; Bob Lutz, GM; Paul Marino, Machinima; Ross Mayfield, Socialtext; Rob McEwen, Goldcorp; Dan McGrath, IBM; Robert Merges, Berkeley University; Steve Mills, IBM; Beth Noveck, NYU; David Pescovitz, Boing Boing; Josh Peterson, 43 Things; Kim Polese, SpikeSource; Judy Rebick, Rabble; Greg Reinacker, NewsGator; Howard Rheingold, Institute for the Future; Philip Rosedale, Second Life; Eric Schmidt, Google; Jonathan Schwartz, Sun Microsystems; David Sifry, Technorati; Jim Stengel, P&G; Robert Stephens, Geek Squad; Phil Stern, yet2.com; Linus Torvalds, Linux; Sophie Vanderbroek, Xerox; Jimmy Wales, Wikipedia; Andy Walter, P&G.

Our close colleagues at New Paradigm and those in our business web did some heavy lifting on this book, and made great contributions throughout. These are:

Paul Artiuch, Robert Barnard, Paul Barter, Pierre-Luc Bisaillon, Nick Bontis, Erik Brynjolfsson, Grant Buckler, Scott Burg, Art Caston, Ann Cavoukian, Charlie Fine, Willem Galle, John Geraci, Corinne Gibas, Bill

Gillies, Rena Granofsky, Peter Haine, Denis Hancock, Dan Herman, Phil Hood, Moritz Kettler, Lisa Hutcheson, Lenni Jabour, Samir Khan, Natalie Klym, Ming Kwan, Del Langdon, Erin Lemon, Alex Lowy, Alan Majer, Darren Meister, Denis O'Leary, Jason Papadimos, Bob Parker, Neil Pasricha, Brendan Peat, Joe Pine, Deepak Ramachandran, Stanley Rodos, Bruce J. Rogow, Hubert Saint-Onge, Heather Shaw, Max Stevens-Guille, Peter Suma, Bob Tapscott, Katharine Tapscott, Tim Warner.

Most important we would like to acknowledge the seminal contribution to this work of several New Paradigm executives. Antoinette Schatz and Jody Stevens—the first two employees of New Paradigm—ensured that the operation functioned. Rahaf Harfoush, our book coordinator extraordinaire, provided stellar support throughout the research and writing process. Mike Dover, vice president of syndicated research, kept the trains running on our research—an enormous challenge. Joan Bigham, executive vice president of strategy and business development, provided invaluable insights throughout and ensured that the work was funded. David Ticoll, executive vice president of research, cocreated a number of profound ideas that are central to this book and has been a close collaborator and partner for two decades. To all we are enormously grateful.

We would also like to thank the Portfolio team at Penguin: Adrian Zackheim, Adrienne Schultz, and Will Weisser for their confidence in us and for their hard work on the manuscript; and our agents Bill Leigh and Wes Neff of the Leigh Bureau for their wise counsel throughout.

Our respective wives, Ana Lopes (Don) and Michelle Allicock (Anthony), provided valuable ideas and advice that repeatedly forced us to make the book accessible to normal human beings. We owe them our highest gratitude.

Having said all this, we as authors take full responsibility for the content of this book, as well as for any errors or omissions.

NOTES

Preface

1. Martin Wroe, "200 Million Heads Are Better Than One, so Join the Crowd." *The Sunday Times* (September 2, 2007).
2. Nicholas G. Carr, "The Ignorance of Crowds." *strategy + business*, issue 47 (Summer 2007).

Introduction

1. The Hypernet Revolution research program was led by Tim Warner under the direction of David Ticoll and Don Tapscott and conducted by our predecessor, Digital 4Sight. The program was sponsored by twenty-five large corporations.
2. Leadership in the Networked Economy, a $2 million effort led by Don Tapscott and Anthony Williams and sponsored by a dozen companies.
3. Information Technology and Competitive Advantage, led by Don Tapscott, David Ticoll, Joan Bigham, and Mike Dover, was funded by twenty-two global corporations and conducted between September 2004 and December 2005.

1. Wikinomics

1. The term "peer production" was coined by Yale professor Yochai Benkler. See Yochai Benkler, "Coase's Penguin, or, Linux and the Nature of the Firm," *Yale Law Journal*, vol. 112, (2002–2003). Throughout the book we use peer production and mass collaboration interchangeably.
2. According to Technorati.com, which monitors registered blogs. The actual number including unregistered blogs is much higher.
3. At the time of writing, Boing Boing was receiving 750,000 visits per day.
4. Jim Giles, "Internet encyclopedias go head to head," *Nature*, vol. 438 no. 531 (December 15, 2005) www.nature.com/news/2005/051212/full/438900a.html.

5. The term "Web 2.0" was coined by O'Reilly vice president Dale Dougherty in 2004. Tim O'Reilly, "What Is Web 2.0?" oreillynet.com (September 30, 2005).

6. For example, as global complexity increases so does the list of challenges we face that are unsolvable by individual organizations acting alone. Tackling global warming. Defeating poverty and disease. Finding new energy sources. Building nano-size supercomputers. Not just sequencing the human genome, but genuinely understanding it. There is simply no end to the requirements or possibilities for innovation. These complex problems demand cross-disciplinary and interorganizational solutions. Even comparatively simple products are becoming more complex. All of this complexity is fueling an increase in the requirement for openness and boundary-spanning collaborations.

7. Don Tapscott and Anthony Williams, "Creating Value in the Age of Transparency," Conference Board (2003). Don Tapscott and David Ticoll, *The Naked Corporation: How the Age of Transparency Will Revolutionize Business* (New York: Simon & Schuster, 2003).

8. National Intelligence Council, "Mapping the Global Future: Report of the National Intelligence Council's 2020 Project" (December 2004).

9. James W. Cortada and David Ticoll, "On Using IT Standards As a Competitive Tool in a Global Economy," New Paradigm's IT & CA Research Program, Big Idea Series (2005).

10. The electric dynamo (a device for converting mechanical energy into electricity) was invented in 1832, but the diffusion of electric motors in U.S. industry was a protracted process, and the anticipated leaps in productivity were long delayed. Much of the delay reflected the fact that both public and private organizations needed time to absorb the new technologies and their impacts. When a new techno-economic paradigm finally matured, we experienced over half a century of sustained growth and productivity during the post–World War II period. See Stanford economist Paul David's work for a fascinating look at how technology figures in recent economic history.

2. The Perfect Storm

1. Don Tapscott, David Ticoll, and Alex Lowy first used this perfect storm analogy in the *E-Company Now* cover story, "Internet Nirvana," March 2001.

2. The Age of Networked Intelligence is an age of promise. It is not simply about the networking of technology but about the networking of humans through technology. It is not an age of smart machines but of humans who, through networks, can combine their intelligence, knowledge, and creativity for breakthroughs in the creation of wealth and social development. It is an age of vast new promise and unimaginable opportunity.

Think about scientific research. In the past, scientists would work with a powerful supercomputer to, say, simulate mechanisms of a biological cell membrane as a way to understand the structure of biological molecules. But as networking permeates the planet, computers everywhere can be marshaled concurrently to attack the problem.

Rather than a single expensive computer supporting a single group of scientists, a global network of computers can be internetworked to support distributed teams of scientists. The network becomes the computer—infinitely more powerful than any single machine. And networked human intelligence is applied to research, thus creating a higher order of thinking, knowledge—and maybe even internetworked consciousness—among people.

The same networking can be applied to business and almost every other aspect of human endeavor—learning, health care, work, entertainment.

Networking can change the intelligence of a business by bringing collective know-how to bear on problem solving and innovation. By dramatically opening up the channels of human communication, is it possible that consciousness can be extended from individuals to organizations? Unconscious organizations, like people, cannot learn. Through becoming conscious, organizations can become able to learn—and that's a precondition for survival. Networked intelligence is the missing link in organizational learning and the conscious organization may be the foundation for the elusive learning organization. And perhaps networked intelligence can be extended beyond organizations to create a broader awakening—social consciousness—in communities, nations, and beyond. (Don Tapscott, *The Digital Economy: Promise and Peril in the Age of Networked Intelligence.* New York: McGraw-Hill, 1996).

3. The idea itself goes back centuries and, most notably in more recent times, to an economist named Friedrich Hayek. Hayek was an outspoken critic of hierarchically driven planning and felt that fellow economists too often took for granted the spectacular, largely self-organizing evolution of the price system and its usefulness in coordinating economic activity. He argued that a decentralized mechanism like price discovery harnesses the dispersed knowledge of individuals to arrive at far better decisions about how to allocate resources than if we were to rely on the individual judgments of experts. Though Hayek's theories, forged in the heyday of socialist thinking, were not always popular, most economists now believe that Hayek was right. Harvard economist Jeffrey Sachs recently stated: "If you ask an economist where's a good place to invest, which industries are going to grow, where the specialization is going to occur, the track record is pretty miserable. Economists don't collect the on-the-ground information, businessmen do." Quoted in "Friedrich August Hayek (1899–1992)," The Concise Encyclopedia of Economics. http://www.econlib.org/library/Enc/bios/Hayek.html.

4. Smarter software, for example, will help eliminate small discrepancies and simple features on services like del.icio.us, such as recommending tags based on common tags that other del.icio.us users have already used to describe the same Web link, and help users converge on common vocabularies for describing content.

5. In the longer term it's foreseeable that tagging could evolve into a new, organic, community-based search function. A tag for a site is like a "vote" (i.e., this site is a good source of information on computer viruses). People who have a knack for identifying quality content can be identified through statistical analysis. The opinions of these people can then be given a greater weight in guiding future searches. Communities could function as a continuously updated representative democracy, as the balance between mathematical algorithms and human opinion are adjusted to create more intelligent search applications.

6. Chicagocrime, for example, is a Web site that injects Chicago crime stats into Google's online maps to create a unique new public service where you type in an address and up pops a map detailing the latest reported crimes in the area. Alternatively, users can specify a crime, such as assault or automated-teller-machine theft, and view a map where those crimes recently occurred.

   Neither Google nor the city of Chicago foresaw this particular application, but the creator of chicagocrime, Adrian Holovaty, did. A small disclaimer on the Web site declares: "This web site is not affiliated with the Chicago Police Department." The openness of the Web makes it a petri dish for exactly this kind of emergent innovation.

7. Don Tapscott. *Growing Up Digital: The Rise of the Net Generation* (New York: McGraw-Hill, 1997).

8. But then the idea that large-scale corporations would replace the cottage industries of Adam Smith's time would have seemed equally foreign to an eighteenth-century Briton. When Adam Smith wrote the defining works of economic theory, capitalism worked like a free market of small businesses and production was mostly small scale. Farmers and craftspeople sold their goods in markets according to local conditions of supply and demand. There was very little production planning, forecasting, or marketing. If you wanted a coat, you engaged a seamstress or tailor to make one. If you needed eggs, you bought them in the market or haggled with a local farmer.

9. Ford Motor Company was the quintessential example of this. Henry Ford had evidently concluded that capitalism required direct control and coordination. Ownership of production processes across the automotive value chain was warranted, even if individual links were thereby shielded from market disciplines.

10. The concept of Coase's law was developed by Tapscott, Ticoll, and Lowy in the book *Digital Capital: Harnessing the Power of Business Webs* (Cambridge, Mass.: Harvard Business School Press, 2000). The first book to discuss the

relevance of Ronald Coase's theories to the Internet was by Don Tapscott, *The Digital Economy: Promise and Peril in the Age of Networked Intelligence* (New York: McGraw-Hill, 1995).

11. Year after year China and India steam ahead in the race to become the world's next economic superpowers. Indeed, the spiraling growth of R&D spending, graduates, full-time researchers, patents, and scientific papers—all of which are on a rapid incline—provide analysts with a daily grind of new statistics. We know, for instance, that for the cost of one engineer in the United States, a company can hire about five engineers in China and eleven in India. Soaring exports, astonishingly low prices, and robust annual growth rates only bolster the feeling of economic inevitability.

12. McKinsey, the business consultancy, estimates that China may face a talent crunch, not a glut, in its bid to transition from manufacturing to services. The demand for labor from just the large foreign-owned companies and joint ventures that now do business in China highlights the problem. From 1998 to 2002, employment in these two categories rose by 12 percent and 23 percent a year, respectively, to about 2.7 million workers. Assuming that 30 percent of these workers must have at least a college degree, and that the labor demands of such companies continue to grow at the same rates, they will have to employ an additional 750,000 graduates from 2003 through 2008. China, we estimate, will produce 1.2 million graduates suitable for employment in world-class service companies during that period. So large foreign multinationals and joint ventures alone will take up to 60 percent of China's suitable graduates before demand from smaller multinationals or Chinese companies even enters the picture. See Diana Farrell and Andrew J. Grant, "China's looming talent shortage," *McKinsey Quarterly*, no. 4 (November 2005).

13. India, for example, has a burgeoning and remarkably successful generic pharmaceuticals industry, where the critical drivers are basic chemistry skills and quality manufacturing at low cost. But there are no breakthrough cures for cancer emerging from Indian pharmaceutical firms just yet.

Likewise, China hasn't produced anything pathbreaking to become the world's leading manufacturer of toys, textiles, or computer chips. Its success factors are low price points, scale, and variety, which the Chinese deliver in abundance. Madhav Bhatkuly quoted in BusinessWeek Online, "Can China and India Innovate?" August 22, 2005.

14. Arnoud de Meyer, deputy dean of INSEAD and author of a recent study of Asian innovation, argues that Asian firms lack requisite skills and experience to take on seasoned innovators with well-honed skills in management, conceptual thinking, and marketing. For example, large mental and geographical distance from the most sophisticated consumer markets, coupled with underdeveloped local markets, means Asian firms often lack sufficient exposure to

trend-setting customers. Managers are accustomed to focusing on cost reduction and process optimization, and seldom think in terms of unique value creation or continuous product innovation. Many executives lack experience in global environments and face challenges in reconfiguring capabilities to adapt to fast-changing markets. And managers typically lack an appreciation for the need to nourish strong brands or capitalize on intellectual property.

15. Similarly, IT and consumer electronics giants like Hewlett-Packard, Motorola, Nokia, and Philips are turning to independent design houses in Asia. Though much of this work could accurately be described as "development" (i.e., tailoring products to local markets), design shops such as HTC, Flextronics, and Wipro Technologies are climbing farther up the food chain. Whereas design shops once focused mainly on optimizing subsystems or components, they increasingly get involved at an earlier stage where they help turn concepts into actual products and services. Outsourcing these capabilities comes with some IP risks, but allowing others to cocreate some of the value is the only way to cut costs and speed time to market.

16. Pete Engardio, "The Future of Outsourcing." *BusinessWeek*, January 30, 2006. www.businessweek.com/magazine/content/06-05/63969401.htm.

3. The Peer Pioneers

1. en.wikipedia.org/w/index.php?title=7_July_2005_London_bombings&action=history.

2. For a pioneering and definitive account of these changes see Yochai Benkler, "Coase's Penguin, or Linux and the Nature of the Firm," *Yale Law Journal*, vol. 112 (2002).

3. Steven Weber, *The Success of Open Source* (Cambridge, Mass.: Harvard University Press, 2004).

4. IBM director of Linux development Daniel Frye estimates that in total more than one thousand serious contributors work on the GNU/Linux kernel, and up to twenty thousand make contributions to the overall operating system. Hundreds of thousands of programmers have made contributions to other open source software projects, many times more than any individual firm could muster. Yochai Benkler estimates that the current crop of one billion people living in affluent countries has between two billion and six billion spare hours among them, every day! If even a small fraction of this creative capacity could be harnessed to produce high-quality information-based goods, the output of these voluntary efforts would dwarf the output of today's knowledge-intensive industries. All it takes is the desire to create and the tools to collaborate, and both of these are increasingly in abundance. Now imagine the productive capacity of billions of people self-selecting for tasks with little regard for organizational, national, cultural, or disciplinary boundaries. *See*

Yochai Benkler, *The Wealth of Networks* (New Haven: Yale University Press, 2006).

5. Jimmy Wales, "The Intelligence of Wikipedia" webcast, Oxford Internet Institute, July 11, 2005, webcast.oii.ox.ac.uk/?view=Webcast&ID=20050711_76.

6. Jim Giles, "Internet encyclopedias go head to head," *Nature*, vol. 438 no. 531 (December 15, 2005) www.nature.com/news/2005/051212/full/438900a.html.

7. It's also true that while Wikipedia scores highly on scientific subjects where the facts are well-documented, it fairs less well on subjects where emotion and ideology can affect the objectivity and accuracy of the article, like history and politics, for example.

8. Robert McHenry, "The Faith-Based Encyclopedia," *Tech Central Station* (November 15, 2005). www.techcentralstation.com/111504A.html.

9. As Stacy Schiff recently pointed out in the *New Yorker*, Wikipedia can aspire to be all-inclusive because there are no physical limits on its size. As a result, Wikipedia covers a breadth and diversity of topics that traditional encyclopedias can't accommodate in their bound pages. "Apparently, no traditional encyclopedia has ever suspected that someone might wonder about Sudoku or about prostitution in China," quips Schiff. "Or, for that matter, about Capgras delusion (the unnerving sensation that an impostor is sitting in for a close relative), the Boston molasses disaster, the Rhinoceros Party of Canada, Bill Gates's house, the forty-five-minute Anglo-Zanzibar War, or Islam in Iceland." (Stacy Schiff, "Can Wikipedia Conquer Expertise?" *The New Yorker*, July 31, 2006.)

10. Jimmy Wales as quoted by Thomas Goetz, "Open Source Everywhere," *Wired* issue 11.11 (November 2003).

11. As for Linus being bad at making predictions, we're reminded of the saying "The future is not something to be predicted, it's something to be achieved." He created a new future, so maybe we should cut him some slack about being skeptical of the genie he helped unleash and how powerful it could become.

4. Ideagoras

1. InnoCentive now operates independently of Lilly and is actively extending its network by partnering with universities in China, India, and Russia. Contract research organizations are also getting involved, many of which see InnoCentive as a great way to utilize their excess capacity and get paid to market their skills.

2. The term "agora" is not heard often these days, but translated from Greek it means marketplace—a place where people can assemble to debate and barter. The Ancient Agora of Athens, for example, was the heart of ancient Athens, the focus of political, commercial, administrative, and social activity, the seat of justice, and the religious and cultural center.

3. Memberships, which start at about ten thousand dollars per year, allow companies to post an unlimited number of offerings on the exchange. After that royalty fees on the sale of products incorporating a technology or patent range from 0.5 percent to about 3 percent. Yet2 collects a finder's fee of 15 percent of the royalties. Some subscribers, including P&G, have also invested in the exchange itself.

4. Quoted in "Internet Tech Transfers at General Electric Industrial Systems," yet2.com (December 3, 2001).

5. Larry Huston and Nabil Sakkab, "Connect and Develop: Inside Procter & Gamble's New Model for Innovation," *Harvard Business Review*, vol. 84, no. 3 (March 2006).

6. Notwithstanding detractors of IT such as Nicholas Carr, whose article in the *Harvard Business Review* ("IT Doesn't Matter," May 2003) caused a considerable brouhaha among readers. For the definitive rebuttal of Carr's thesis, see Don Tapscott, "Rethinking Information Technology and Competitive Advantage— the Debate," available at www.newparadigm.com.

7. Quoted in "Internet Tech Transfers at General Electric Industrial Systems," yet2.com (December 3, 2001).

8. Quoted in Alex Wood and Alex Scott, "Licensing activity is on the rise." *Chemical Week* (March 24, 2004). www.chemweek.com/inc/articles/t/2004/03/24/006.html.

5. The Prosumers

1. Robert Hof, "My Virtual Life," *BusinessWeek* (May 1, 2006) www.businessweek.com/magazine/content/06_18/63982001.html.

2. Ibid.

3. Alvin Toffler first coined the term "prosumer" in his book *The Third Wave* (New York: Bantam Books, 1980).

4. In fact, much of the roar over customer centricity emerged when the Web's first superstars placed customers at the core of their operations. When you shop on Amazon, for example, you select, purchase, and schedule the delivery of products yourself. You also help sell products, by reviewing them, creating favorites lists, and contributing data regarding your own purchases for Amazon's associative modeling system, which, in turn, generates personalized recommendations for all Amazon customers. Likewise, eBay may provide the context for its online exchange and helps broker trust among participants, but you create most of the "content" when you buy and sell goods with fellow traders.

5. In his book *Democratizing Innovation*, Professor Eric von Hippel of MIT tells the story of how mountain bikers (an exemplary lead user group) changed the face of the bicycle industry. Well before genuine mountain bikes emerged on

the market in the mid-1980s, cycling enthusiasts who liked to ride on rugged terrains, perform stunts, and endure difficult weather conditions were modifying their gear to suit the extreme conditions. Early lead users typically assembled custom units with strong bike frames, balloon tires, and powerful drum brakes designed for motorcycles. But many users went further. One trickster developed his own armor and protective clothing. Another back-country biker invented a harness to carry his bike up steep mountains and dangle it over cliffs. Yet another embedded metal studs in his tires for biking on ice. All of these innovations are now mainstream fare for mountain bikers today.

Professor von Hippel notes that mountain biking grew to over half a million participants before manufacturers started to make bicycles suited for these users. For over a decade the niche market for mountain bikes remained the preserve of specialized producers and custom bike shops. Today sales of mountain bikes are big business (about $60 billion) and represent roughly 65 percent of the U.S. bicycle market.

6. Deep customer engagement in the gaming industry is yet another signpost for an era of customer-driven innovation. Electronic Arts (EA) ships programming tools to its customers, and then posts their modifications to the games online. Many customer modifications get added to EA's final products. Westwood Studios (now owned by EA) has been sharing game development tools for eight years. Since 1999 it has worked closely with customers and external software developers to cocreate its games.

7. "Confused in Calcutta," August 21, 2006.

8. The story of how a clever hacker from Frankfurt, Germany, reverse engineered the iPod's operating system and ported a Linux install is remarkable in and of itself. It took four months of excruciating effort and lots of line-by-line analysis of the software code (for the details, check Wikipedia).

9. In a bid to turn ordinary players into console game makers, for example, Microsoft introduced a test version of Xbox 360 software in August 2006 that it says will considerably simplify the creation of basic games for its console—the very thing that Sony most fears. Microsoft would rather delight its customers with the option to customize, build, and share their own games, however, than waste energy defending a monopoly over game development on its console. More and better games will help sell consoles and, in the end, Microsoft's shrewd move to increase customer participation may help blunt the impact of Sony's PlayStation 3 (due to be released at the end of 2006).

10. Renee Graham, "Will ruling on samples chill rap?" *Boston Globe* (September 14, 2004).

11. Hip-hop stalwarts worry that sonic masterpieces such as the Beastie Boys' "Paul's Boutique," released in 1989, or Public Enemy's "It Takes a Nation of Millions to Hold Us Back" are a thing of the past.

12. Sarah Frere-Jones, "The New Math of Mashups," *The New Yorker* (January 10, 2005) www.newyorker.com/critics/music/?050110crmu_music.

13. The Creative Commons was launched in December 2002, and within the first year there were over a million Web-based projects using some form of Creative Commons licensing. After two years there were four million, and by January 2006 there were over fifty million, which doesn't include seven million images on flickr that are licensed under the Creative Commons.

14. The film, which premiered at the Sundance Film Festival in January 2005, was scooped up by indie distributor ThinkFilm for a cool seven figures, and then released on July 7, 2006, to an enthusiastic worldwide audience. Total cost of production for the Beastie Boys: about $1.2 million.

15. Jeff Jarvis, "Can you Digg what is happening to journalism?" *The Guardian* (February 27, 2006).

16. Ibid.

6. The New Alexandrians

1. Kelly predicts that when fully digitized the whole lot could be compressed (at current technological rates) onto fifty petabyte hard disks. "Today you need a building about the size of a small-town library to house fifty petabytes," he says. "With tomorrow's technology it will all fit onto your iPod." And by the time that happens, in five to ten years, we humans will have more than doubled our stock of knowledge. (Kevin Kelly, "Scan This Book!" *New York Times*, May 14, 2006).

2. Kevin Kelly, "Speculations on the Future of Science," *Edge*, vol. 179 (April 7, 2006).

3. Quoted in Bob Garfield, "Inside the New World of Listenomics." AdAge .com (October 11, 2005).

4. See Joel Mokyr, *The Gifts of Athena* (Princeton: Princeton University Press, 2002) for a book-length treatment of this concept.

5. The ideas of Francis Bacon and Isaac Newton, which defined the scientific method, set the tone for much of what would follow in the century. Bacon and Newton believed that true science called for axiomatic proof to be fused with physical observation in a coherent system of verifiable predictions. For scientific theories and predictions to be verifiable, science needed to be open.

6. Joel Mokyr provides a fantastic history of how the institutions that created bridges between the realm of science and the realm of private enterprise became more robust with time. Scientific debating societies, for example, were augmented by public universities, polytechnic schools, publicly funded research institutes, museums, agricultural research stations, and research departments in

large corporations and financial institutions. Textbooks, professional journals, encyclopedias, and trade publications appeared in every field and made it easier to look things up. The growing abundance of experts meant that anyone who needed some piece of useful knowledge could find someone who knew, or who knew someone who knew.

7. The average number of authors per scientific paper is up too, increasing steadily over the past sixty years from an average of slightly over 1 to averages of 2.22 in computer science, 2.66 for condensed-matter physics, 3.35 for astrophysics, 3.75 for biomedicine, and 8.96 authors for high-energy physics. A growing number of papers have between 200 and 500 authors, and the highest-ranking paper in the study had an astonishing 1,681 authors. See: M. E. J. Newman, "Who is the best connected scientist? A study of scientific co-authorship networks," *Working Paper*, Santa Fe Institute (2000).

8. At the Sloan Digital Sky Survey, for example, hundreds of researchers from fifty institutions worldwide are harnessing the power of ten thousand computers and over fifteen terabytes of data to search for and analyze millions of nearby planets and stars. The free and open exchange of information and ideas will provide scientists with an unprecedented map of the universe in a fraction of the time it would have taken using conventional methods. *See* "Let data speak to data," *Nature*, vol. 438, no. 531 (December 1, 2005).

9. Quoted in "From the Los Alamos Preprint Archive to the arXiv: An Interview with Paul Ginsparg," *Science Editor* 25, no. 2 (March–April 2002): 43.

10. Quoted in "Royal Society: Rent-seeking is more important than science," *Boing Boing* (November 25, 2005) (http://www.boingboing.net/2005/11/25/royal_society_rentse.html).

11. "Let data speak to data," *Nature*, vol. 438, no. 531 (December 1, 2005).

12. Few scientists, for example, think the paper-based approach to scientific publishing is likely to disappear any time soon, not least because of its centrality to the system of academic reward and evaluation. "Everything eventually ends up in a peer-reviewed journal," says Paul Camp, "because that is what counts for tenure and promotion." Paul Myers, a biologist at the University of Minnesota, a blogger on Pharyngula, thinks that the standard scientific paper will be complemented, not replaced, by more dynamic and collaborative forms of communication. Myers describes conventional scientific papers as "static" and "very limited." In the same breath he emphasizes that "the standard scientific paper is irreplaceable as a fixed, archivable document that defines a checkpoint in a body of work." Quoted in Declan Butler, "Science in the web age: Joint efforts." *Nature*, vol. 438, no. 531 (December 1, 2005).

13. "Let data speak to data," *Nature*, vol. 438, no. 531 (December 1, 2005).

14. National Institutes of Health press release, August 22, 2005.

15. Merck press release, February 10, 1995.

16. Quoted in Robert Langreth, Michael Waldholz, and Stephen D. Moore, "Drug Firms Discuss Linking Up to Pursue Disease-Causing Genes." *Wall Street Journal* (March 4, 1999).

17. Ibid.

18. Ibid.

19. Ibid.

20. The trends are the same around the world: The annual number of new active substances approved in major markets fell by 50 percent over the 1990s, while private sector pharmaceutical R&D spending tripled, to $47 billion.

21. MIT press release announcing the creation of the Center for Biomedical Innovation, "MIT Launches Center for Biomedical Innovation." April 29, 2005.

22. David Tennenhouse has since moved on to become CEO of A9, Amazon's search engine division.

23. This quote and all further quotes from: David Tennenhouse, "Intel's Open Collaboration Model Industry-University Partnerships." *Research Technology Management*, vol. 47, no. 4 (July–August 2004).

24. Incremental innovation is not just a property of the semiconductor industry. In process- and scale-intensive industries such as automotives, steel, and consumer goods, firms have been chasing Chilton's law, according to which doubling plant capacity only increases investment cost by two-thirds. The arrival of catalytic refining in the oil industry, combined with improved capabilities to profile demand with oil derivatives, lead to successive increases in the yield of gasoline. The "green revolution" introduced new families of agricultural machinery and multiple petrochemical innovations in fertilizers, herbicides, and pesticides, which together drove predictable improvements in the productivity of agricultural systems. Although Viagra doesn't get smaller, smarter, faster, or cheaper every year, we do see regular improvements in the safety and efficacy of existing drug classes, along with new formulations, dosing schedules, and administration options that significantly increase patient compliance.

25. In one example, Intel's work with UC Berkeley on a new sensor net tool kit called "TASK" (Tiny Application Sensor Kit) has been widely shared among university researchers outside of the project. Today the TASK tool kit is creating a common infrastructure for researchers in the sensor network space. Thousands of researchers around the world are now identifying new applications for the technology.

26. Until quite recently the not-for-profit sector pursued largely taxpayer-supported, curiosity-driven research inside government labs, universities, and research institutes. Researchers concentrated largely on basic science, published early and extensively, and filed very few patents. In a world of dwindling public dollars and heightened competition for funding, some of these norms

are changing. Increasing resource requirements and societal pressure to justify budgets are pushing public institutions to focus more on applied research, patent discoveries, and licensing revenue. Universities and other government-funded institutions have become not merely more tolerant of "off-campus" commercial activity, but active encouragers of it.

Licensing inventions has been a boon for research-intensive universities, but a growing number of people in industry think intellectual property rights often get in the way of effective industry-university collaboration. The problem lies in the process of valuing the intellectual property developed in industry-university collaborations, which can add significant delays to highly time-sensitive commercialization strategies. In the worst case, delays could result in the technology being developed too late to be of financial benefit to anyone. The added costs and delays associated with bargaining over IP have been driving a number of U.S. firms to contract with universities overseas, where deals can be sealed in weeks instead of the months or years typical of universities in the United States.

27. Interview with authors.

7. Platforms for Participation

1. Robert D. Hof, "The Power of Us," *BusinessWeek* (June 20, 2005) www .businessweek.com/magazine/content/05-25/63938601.htm.

2. Other grassroots initiatives soon emerged to complement the PeopleFinder efforts. Two Web designers in Utah, for example, launched Katrinahousing to connect those displaced by Katrina with people who could house them. News reports suggest that five thousand people had found homes using the free service within two weeks of the disaster. And, for those who lacked Internet access (which many naturally did), volunteers and tech companies constructed makeshift wireless networks, set up computers at emergency shelters, and distributed Internet-ready phones so that evacuees could make use of these Web services.

3. Steve Berlin Johnson, "I'm Looking for Uncle John," *Discover*, vol. 26 no. 12 (December 2005).

4. Ryan Singel, "Map Hacks on Crack," *Wired* (July 2, 2005) www.wired.com/ news/technology/0.68071-0.htm.

5. "Dyke to open up BBC archive," BBC News (August 24, 2003).

6. Robert D. Hof, "The Power of Us," *BusinessWeek* (June 20, 2005) www .businessweek.com/magazine/content/05-25/63938601.htm.

7. In fact, to give you a sense of how freewheeling all of this is, Jeff Barr, Amazon's Web services evangelist, hadn't even heard of these particular examples when we called him up to ask about them.

8. Jim Willis blog, Rhode Island GovTracker Services, Government Open Code Collaborative (June 30, 2005) www.gocc.gov/members/sjwillis/weblog_storage/blog_07956.

9. This initial project, which was led by Neal Richman, director of NKCA and associate director of UCLA's Advanced Policy Institute, found that the best predictor of housing abandonment is tax delinquency. The researchers involved with the NKCA project began using tax data to look for a characteristic pattern in housing serving low-income residents. Property tax delinquency, they found, is often followed by building code violations and tenant complaints and, eventually, abandonment.

After coming across a similar early warning system in Chicago, graduate student Daniel Krouk proposed that an interactive database approach could yield a powerful policy research tool with wider use and accessibility than a traditional research study. An online project prototype was developed and presented to the City of Los Angeles Housing Department, which provided initial funding. The Fannie Mae Foundation and the U.S. Department of Commerce Telecommunications and Information Infrastructure Assistance Program were also approached, and subsequently provided major financial assistance. A condition of funding was that the UCLA team find a home for NKCA in a nonprofit organization so that the tool would be directly accountable to an active constituency.

10. Jeff Jarvis, "Who owns the wisdom of the crowd?" BuzzMachine (October 26, 2005).

11. Ibid.

12. YouTube founder Chad Hurley, for example, admits that despite its rocketing popularity, the site is still not profitable. However, with advertising picking up and negotiations now underway with major Hollywood studios, Hurley claims that profitability is not far off. (John Boudreau, "YouTube Strategy Sticks to Clips," *Mercury News*, June 27, 2006.)

8. The Global Plant Floor

1. Gershenfeld and colleagues believe the immediate promise of personal fabrication lies in developing countries, where aspiring inventors and entrepreneurs often lack the resources to implement their ideas. Investing in the full-fledged physical infrastructure to build physical goods is expensive and risky. Scarce capital and steep costs thwart the ability of local entrepreneurs to develop homegrown manufacturing businesses, and this in turn keeps developing countries dependent on importing expensive technology from outside.

Fab labs, on the other hand, provide entrepreneurs with the low-cost tools to translate back-of-the-envelope designs into working prototypes. MIT fab labs are equipped with the same capabilities, so inventors can easily share digital designs and fabricated solutions between labs in true open source fashion. Communities can even link up with a worldwide roster of engineers willing to swap design solutions and help solve problems. With the help of a growing network of microcredit agencies to provide financing, some of these prototypes will spur local business and in the long run help fuel their countries' development aspirations.

2. Samuel Palmisano, "The Evolving Global Enterprise," *Foreign Affairs* vol. 85, no. 3 (May/June 2006).

3. As we explained in the opening chapters, this is all part of a profound economic revolution in which so-called vertically integrated firms are unbundling and refocusing on strategic and unique core capabilities. This does not mean that vertical integration never makes sense, or that firms should not expand or evolve their core to create new value or penetrate new markets. It does mean that decisions about what falls inside and outside of corporate boundaries have become critical to the effective orchestration of resources and, hence, competitive advantage.

4. Palmisano, "The Evolving Global Enterprise."

5. The process starts with a reverse-engineering perspective. Assemblers choose a focal or target product to knock off, often in consultation with suppliers. Then they develop product architectures so that supplying firms can develop many components independently and in parallel with the assemblers.

   The major motorcycle assemblers take a lead role in orchestrating the design and assembly process. They brand and market the end product but have varying responsibilities for design, sourcing, and manufacturing. Most are content to set parameters like weight and size, leaving detailed design work to the component suppliers.

6. Quoted in Arnie Williams, "Boeing's 7E7 Project Pushes PLM Boundaries: Digital behavior modeling from concept through lifecycle," *CADalyst* (April 2004).

7. Ibid.

8. Quoted from a speech delivered to the SAE World Congress, Detroit, Michigan, April 13, 2005.

9. Ibid.

10. Ibid.

11. Ibid.

12. Not surprisingly, suppliers are equally enthusiastic about the collaborative approach to innovation and manufacturing. Working side-by-side with BMW engineers enables the employees at suppliers like Friedrichshafen to hone their capabilities and learn new skills on the job. Suppliers get the job done more quickly and to a higher standard, as BMW's concern for product integrity and quality spreads throughout the supply chain.

13. John Hagel and John Seely Brown, *The Only Sustainable Edge* (Boston: Harvard Business School Press, 2005).

9. The Wiki Workplace

1. To check out their look try browsing through Geek Squad's twelve hundred photos on flickr (flickr.com/groups/geeksquad).
2. As for systems, Stephens spent the nineties building up his own homegrown dispatch systems and administration software in order to save money. But his hard work eventually spawned a well-oiled services organization that delivers support in your home, in the store, over the phone, or online. Most of this software is still in place today. "We built it for thirteen thousand dollars," says Stephens, "and I won't say how much money Best Buy has spent trying to replace it." In the end they ended up scaling-up Stephens's solutions instead.

   Geek Squad's business model is unique in several ways. Top of the list for most customers is the company's policy of charging a flat rate for services, when most repair services charge by the hour. This keeps it simple for customers by taking the guesswork out of pricing. In some cases flat-rate pricing means some people pay more (a home visit for configuring a broadband connection costs $159; cleaning out spyware, $129). "But people are willing to pay for simplification," says Stephens. At the same time, flat rates cut down on paperwork and eliminate the need to monitor staff for fraud.
3. Stephens claims the potential for a great marriage presented itself early on. "Their customers were already coming to me," he said. "And I'd tell them, 'Mrs. Johnson, I can fix your laptop, and it's going to cost $250, but you've got a warranty with Best Buy.'" Best Buy could fix the laptop for free, but most customers were unhappy with the service. Stephens says that customers like Mrs. Johnson would come back to him complaining that Best Buy didn't fix it right, and that they didn't want to wait three weeks to get it back. "They would happily pay $250 and get it done in three hours," said Stephens.

   Best Buy and Geek Squad started courting in 2000 after Stephens delivered a winning pitch. The formal acquisition in 2002 led to a number of Geek Squad tests runs in stores around the country.
4. For example, Stephens says, "we hire people who spend their spare time playing with technology, which is a form of training that we don't have to pay for." "To be honest," he continues, "our most effective form of training is the Best Buy employee discount."
5. The new organization was given many names. Peter Drucker called it the "networked organization" (*Harvard Business Review*, January–February 1988). He referred to middle management as "the boosters of faint signals that pass for communication in the pre-information organization." Peter Senge coined

the "learning organization" (*The Fifth Discipline*, New York: Doubleday, 1990). For Peter Keene, it was the "relational organization" (*Shaping the Future: Business Design Through Information Technology*, Cambridge, Mass.: Harvard Business School Press, 1991). Tom Peters called it the "crazy organization" (*The Tom Peters Seminar: Crazy Times Call for Crazy Organizations*, New York: Vintage Books, 1994). For D. Quinn Mills it was a "cluster organization" (*Rebirth of the Corporation*, New York: John Wiley and Sons, 1991). Charles Savage called it "human networking" (*5th Generation Management: Integrating Enterprises Through Human Networking*, Burlington, Mass.: Digital Press, 1990). For James Brian Quinn, the "intelligent enterprise" (*Intelligent Enterprise: A Knowledge and Service Based Paradigm for Industry*, New York: Free Press, 1992). And of course we'd be remiss not to mention Don Tapscott's books and writings since 1982 about internetworked enterprises.

6. These N-Geners are not only enthusiastic adopters of technology, they are comfortable with the pace of technological change and optimistic about the role that technology will play in improving our world. A recent MIT survey of American teenagers found that a large proportion of this demographic think gasoline-powered automobiles, landline telephones, compact discs, and desktop computers are headed toward the technology scrap heap as soon as the year 2015. Well over 80 percent believe that technology will provide clean water, end world hunger, eradicate disease, reduce pollution, and contribute to energy conservation well within their lifetime.

7. Quoted in Chris Taylor, "It's a Wiki Wiki World," *Time*, vol. 165 no. 23 (June 6, 2005).

8. Not all employees in the R&D group have taken naturally to the process. Many employees have been with Xerox for decades and are not accustomed to putting their thoughts out on a wiki, where they might be openly criticized. But according to Vandebroek, Xerox's recent intake of younger engineers and scientists really get it. "These people are so amazing," she says. "They are much more open, much more outspoken, and they love these collaborative tools."

9. Management consultancies that specialize in workplace design typically don't help—they force everything into linear processes that can be easily centralized, regularized, analyzed, and controlled by senior managers in the organization.

10. Ironically, we spent much of the last century perfecting decision support systems to get useful information to a centralized place so that experts could make decisions. Much of this effort was misplaced. We ought to have been finding a way to collect the dispersed knowledge of individuals so that the aggregate insights of large groups could be harnessed.

11. Many of the same companies are using internal markets to allocate scarce dollars to R&D projects. The collective hunches of the team are often better at picking winners than the sole discretion of the unit's director. Plus, markets

can track opinions in real time so that judgments can be updated as projects mature and new conditions unfold in the marketplace.

12. Users start with credited accounts and a list of available processors, their current workloads, and a price schedule. Users place bids for various processors, using a sliding price dial that looks like a volume control. If the deadline of one task suddenly moves forward, the user can up his bid and immediately get more processor cycles. As users consume cycles, the software deducts credits from their account.

Huberman's system even includes a "truth-telling" function within the reservation system that helps others determine how likely those reservations are to be kept. Employees can choose a low-cost reservation with a high penalty for cancellation or pay a higher price upfront with minimal penalties for opting out later. Choices about how much employees are willing to pay and how much penalty they will accept reveals a lot about how likely they are to use the reservation, creating an incentive for employees to take only what they need.

13. Schwartz also insists that all leaders in the organization—which by Schwartz's definition means anyone with the title of vice president or director or higher—are responsible for reading employee blogs. "Let me tell you," he says, "I go read those blogs and I see a Sun employee or somebody out in the community, because increasingly you can't tell the difference, and they are criticizing one of our products. And I turn around to the executive who's paid a tremendous amount of money, and I say, why aren't you fixing this problem?"

14. Despite Sun's successes, few other organizations have been so bold. A survey conducted by Socialtext and *Wired* magazine in June 2006 found twenty-nine (that's only 5.8 percent) of Fortune 500 companies are blogging, and most of those companies are in the tech sector. Another tech company that no longer speaks with one corporate voice is IBM. They recently launched a program to encourage every member of its 320,000 strong staff to start blogging and podcasting. Many employees are experts in their fields, and the company thinks they're often the best evangelists for IBM's products and technologies. Over fifteen thousand employees have already registered internal blogs, and over two thousand are actively publishing their thoughts externally.

15. John Seely Brown and Paul Duguid, *The Social Life of Information* (Cambridge, Mass.: Harvard Business School Press, 2002).

10. Collaborative Minds

1. Andrew Keen, "Web 2.0." *The Weekly Standard* (February 15, 2006).
2. "Weighing the Merits of the New Webocracy." *San Francisco Chronicle* (October 15, 2006).

3. Jim Giles, "Internet Encyclopaedias Go Head to Head." *Nature*, vol. 438, no. 900–901 (December 15, 2005).

4. Lawrence Lessig, "Keen's *The Cult of the Amateur:* BRILLIANT!" (May 31, 2007)   http://www.lessig.org/blog/2007/05/keens_the_cult_of_the_amateur .html.

5. "New Program Color-codes Text in Wikipedia Entries to Indicate Trustworthiness." Press release, University of California at Santa Cruz, August 2, 2007.

6. Interview with Scott Hervey, conducted by Anthony Williams, May 24, 2007.

7. Victor Keegan, "Let's Dance to a New Tune." *Guardian* (November 23, 2006) guardian.co.uk.

8. Interview with the authors, February 2006.

9. Interview with Terry McBride, conducted by Anthony Williams, July 4, 2007.

10. Juliette Garside, "Embrace Digital or Die, EMI Told." *The Telegraph* (September 10, 2007).

11. Interview with Fred von Lohmann, senior intellectual property attorney, Electronic Frontier Foundation, conducted by Anthony Williams, June 11, 2007.

12. Ibid.

13. Clayton M. Christensen, *The Innovator's Dilemma: When New Technologies Cause Great Firms to Fail* (Cambridge, Mass: Harvard Business School Press, 1997).

14. Garside, "Embrace Digital or Die."

15. "User-Generated Content Is Top Threat to Media and Entertainment Industry," *Accenture* (April 16, 2007).

16. Interview with the authors, March 2006.

17. Ibid.

18. Don Tapscott and Art Caston, *Paradigm Shift: The New Promise of Information Technology* (New York: McGraw-Hill, 1993).

19. Interview with the authors, February 2006.

11. Enterprise 2.0

1. "Team Unlocks Genetic Basis of Type 2 Diabetes." MIT, February 21, 2007.

2. Stephen Pinnock, "Pharma goes open access," *The Scientist* (February 26, 2007).

3. Interview with authors, April 2006.

4. Lionel Menchaca, "Direct2Dell One Year Later." July 14, 2007. Direct2Dell blog, http://direct2dell.com/one2one/archive/2007/07/14/20884.aspx.

5. Interview with Don Tapscott, May 2006.

6. Interview with authors, February 2006.

7. Dan Bricklin, "The Cornucopia of the Commons." www.danbricklin.com/ cornucopia.htm.

8. Robert D. Hof, "The Power of Us." *BusinessWeek* (June 20, 2005), 82.

9. Interview with Derek Elley, conducted by Anthony Williams, June 2007.

10. C. F. Kurtz and D. J. Snowden, "The New Dynamics of Strategy: Sensemaking in a Complex and Complicated World." *IBM Systems Journal*, vol. 42, no. 3 (2003), p. 464.

11. Joel Cawley quote, interview with the authors, February 2006.

INDEX